The United States Air Force and the Culture of Innovation

1945–1965

Stephen B. Johnson

**Air Force History and
Museums Program**
Washington, D. C.
2002

For sale by the Superintendent of Documents, U.S. Government Printing Office
Internet: bookstore.gpo.gov   Phone: toll free (866) 512-1800;   DC area (202) 512-1800
Fax: (202) 512-2250 Mail: Stop SSOP, Washington, DC 20402-0001

ISBN 0-16-051086-4

Library of Congress Cataloging-in-Publication Data

Johnson, Stephen B., 1959-
    The United States Air Force and the culture of innovation, 1945-1965
/ Stephen B. Johnson.
         p. cm.
    Includes bibliographical references and index.
    1. Aeronautics, Military—Research—United States—History—20th
century. 2. Ballistic missiles—United States—Design and
construction—Management. 3. United States. Air Force—Data
processing—Management. I. Title.
UG643 .J64 2002
358.4'07'0973—dc21

2002003230

Cover Photo

      Original photograph taken by Bud Silloway (CMSgt
Retired), digitally enhanced by Joy Dorman, and adapted
from the version found on-line at WWW.SAGESITE.INFO.
Published and adapted with permission. All Rights Reserved
by on-line Web Site.

NOTE:

To Diane

Foreword

Professor Stephen B. Johnson demonstrates in fine detail how the application of systems management by the United States Air Force to its ballistic missiles and computer programs not only produced critical new weapons, but also benefited American industry. Systems management harmonized the disparate goals of four interest groups. For the military it brought rapid technological progress; for scientists, new products; for engineers, dependability; and for managers, predictable cost. The process evolved, beginning shortly after the end of World War II, when Gen. Henry H. "Hap" Arnold directed that the Army Air Forces (later the U.S. Air Force) continue its wartime collaboration with the scientific community. This started as a voluntary association, with the establishment of the Scientific Advisory Board and Project RAND. In the early 1950s, the Air Force reorganized its research and development (R&D) function with the creation of Air Research and Development Command (ARDC) and the Air Staff's office of deputy chief of staff for development (DCS/D), which were both aimed at controlling the scientists.

The systems management approach evolved out of a jurisdictional conflict between ARDC and its rival, Air Materiel Command (AMC). The latter controlled R&D finances and was determined not to relinquish its prerogatives. Of course, ARDC argued that this was a case of having responsibility without the requisite authority. At first represented by Gen. Bernard A. Schriever's ballistic missiles program, ARDC bypassed traditional organizational structures. Schriever's Western Development Division (WDD), located at Inglewood, California, made its case, based upon the Soviet Union's nuclear threat, to engage in the race to develop long-range ballistic missiles.

Ultimately, Schriever's new project management and weapons systems procedures—concurrency—produced a family of missile and space vehicles. However, in bypassing administrative red tape, this development also eliminated some necessary checks and balances that led to a series of flight test failures and cost overruns.

Closely related to the missiles program was the air defense effort, centered at the Massachusetts Institute of Technology (MIT) in Boston. Dr. Jay Forrester's Project Whirlwind evolved into large-scale, real-time computers. Again, as with the missiles program, once the Cold War waned, the government's emphasis shifted to cost control.

The USAF and the Culture of Innovation

When Schriever assumed command of ARDC, he transplanted his successful Inglewood model to all major weapons systems acquisition. Ironically, in the early 1960s, Secretary of Defense Robert S. McNamara appropriated Schriever's procedures, using them to wield ever greater centralized control.

Dr. Johnson shows that Air Force procedures were not only highly successful in terms of meeting the challenges of the Cold War, but also that their adoption by American industry propelled the nation to international prominence in aerospace and computing. Finally, he argues that while aerospace had experienced somewhat more difficulty adapting to consumer products than did the computer industry, the full implications of systems management were yet to be seen by the end of the Cold War.

RICHARD P. HALLION
Air Force Historian

Preface

The U.S. aerospace and computing industries share a common history, and that history may help explain why those industries have been the most successful industries in global economic competition from the 1950s until the present time. Common to both industries is their genesis in military programs, largely funded by the Air Force. The narrative here cannot confirm or disprove the hypothesis that this common heritage was an important factor in their later successes, but it begins the process needed to assess its validity—that is, historical investigation of those common roots in the Air Force's efforts to develop strategic offensive and defensive systems in the 1950s. To determine whether my instinct is true, many others will have to investigate the economic environment, the effects of military and industrial interactions, and the long-term consequences of systems management as formulated and adopted in the 1960s.

After completing the research for this book, I believe that systems management was one of the significant elements that gave these industries an edge, and perhaps it continues to do so. In particular, I found that configuration control and configuration management had an unexpected importance in the process of innovation. Although the techniques of systems management often are criticized as being overly bureaucratic, configuration control provides a critical coordination mechanism necessary for virtually any development project. Because both the aerospace and computing industries develop highly complex products, the emphasis of systems management on up-front planning, simplification of interfaces, and tight coordination is of fundamental importance. It is my hope that the present work will help dispel a few myths regarding systems management's alleged deadening effect on research and development.

I would like to thank a number of people who helped me with the research for this book. Some of the material, particularly on ballistic missile management, comes from my dissertation research at the University of Minnesota. John Lonnquest has been a good friend and a great resource, first introducing me to the importance of Bernard Schriever's managerial innovations. My dissertation adviser Arthur Norberg was crucial in helping me set a straight course in that work, and Robert Seidel was invaluable in teaching me to write!

The majority of the material for this work, however, is new. My graduate assistant Phil Smith was a great help as he learned my approach to this research and served as my eyes and hands in visiting archives in the Boston area and at Maxwell AFB. Both he and Eric King extracted information from the materials collected and summarized them electronically, a process that sped my work along. I am grateful to them both.

Archivists and historians at the various sites helped me tremendously. David Baldwin at MITRE was the greatest help. He went far beyond duty in tracking down materials and spending time with me while Phil and I sifted through boxes and folders of materials. While I did research for this book in Boston, Randall Bergman of the Defense Technical Information Center at Hanscom Library helped me find some systems management studies I would never have located, and Hanscom historian Ruth Liebowitz got me started on the right track. At Lincoln Laboratory, Roger Sudbury expedited my clearance into their system, and Tamara Granovsky helped Phil and me acquire materials. The Air Force Historical Research Agency staff at Maxwell AFB were uniformly professional and helpful. Rick Sturdevant at the Space Command History Office in Colorado Springs tracked down a number of materials on the North American Air Defense Command and later air defense efforts. I had expected to write more on that topic here, but leaving it for some separate articles makes for a more focused story. Harry Waldron at the Space and Missile Center in El Segundo located some ballistic missile materials and, most importantly, gave me a copy of the shipping records for materials sent to Maxwell, without which I would not have gotten very far.

I also would like to thank Cargill Hall, Richard Davis, and Priscilla Jones at the Air Force History Support Office, Bolling AFB, for their help during my contract—one that enabled me to travel and support a graduate assistantship—and for putting up with some of the unexpected (at least to me) delays. The work turned out to be significantly more extensive than I originally proposed and that led to a five-month delay. I hope the finished book was worth the extra effort!

Finally, I thank my wife Diane and my sons Casey and Travis for being patient with me in the course of this project. They suffered from my overzealous attention to the book over more nights, weekends, and breaks than I care to remember. For their sake, too, I hope that the end result is valuable.

The Author

Stephen B. Johnson earned a Ph.D. in the history of science and technology from the University of Minnesota and is currently an associate professor in the space studies department at the University of North Dakota. The present work is related to his general history of systems management in the Air Force, the National Aeronautics and Space Administration, and the European Space Agency now under review by Johns Hopkins University Press. Prior to taking his present academic post, Johnson was an engineer for Northrop and Martin Marietta and for the University of Cincinnati's Space Engineering Research Center. He was also vice president of engineering for Dependable Systems International and associate director of the Charles Babbage Institute for Computer History at the University of Minnesota.

Contents

The United States Air Force and the Culture of Innovation

1945–1965

Chapter 1

Complexity and the Organization of Research and Development

We are in a technological race with the enemy. The time scale is incredibly compressed. The outcome may decide whether our form of government will survive. Therefore, it is important for us to explore whether it is possible to speed up our technology. Can we for example plan and actually *schedule* inventions? I believe this can be done in most instances, provided we are willing to pay the price and make no mistake about it, the price is high.

Col. Norair M. Lulejian, 1962[1]

From the 1950s through the 1990s, the aerospace and computing industries in the United States were omnipresent symbols of the nation's technological superiority. Whereas many other U.S. industries, such as the once-dominant automobile, steel, and consumer electronics industries, have lost their leading positions to more nimble foreign competition, the aerospace and computing industries largely have maintained their technological and economic edge. Why have they remained successful when so many others have faltered?

Common complaints in the aerospace and computing industries center on how difficult it is to "cross the bridge" from research to development and production, and to work with an overly bureaucratic government. Hearing those complaints, the casual listener and jaded insider believe that the industries are in terrible shape and need drastic reform, particularly in their interactions with the government. But if things really are so bad, how have these industries done so well by one measure that weighs heavily in market success—economic competition against foreign industries?

One simple answer is that these industries received a tremendous amount of government money, and in such circumstances anyone is bound to succeed. If money is the primary factor, however, why have other government-subsidized industries failed? Furthermore, the steel or auto industries did not lack the funds to vie with foreign competitors. The fact is that substantial infusions of capital alone do not explain the difference in success rates.

Recent scholarship has focused on the importance of "organizational learning" as a critical element in organization development and achievement. Organizations that succeed and remain successful usually are those that learn faster and that quickly translate new knowledge into innovative products. Applying this theory, we can see that noncompetitive industries failed to learn the lessons taught them by consumers, competition, or suppliers, or they failed to apply those lessons to produce marketable products. A common difficulty in industry is the inability to move ideas generated in marketing or research departments from the drawing board into the marketplace. Stated another way, the links are weak from research (whether technical or market research) to development and from production to operations, and the chances of failure from one step to the next are frustratingly large. Those organizations or industries that succeed in world competition must have better ways of translating research into products than does their competition.[2]

From that viewpoint, the aerospace and computing industries learned to translate market and technical research effectively into innovative products. How did they learn to do these things? Although the histories of the two industries have significant differences, during the critical period of the 1950s—when the aircraft industry became the aerospace industry, and the office machine industry became the computing industry—both were heavily influenced by the United States Air Force (USAF). To be specific, both industries interacted with the USAF to develop the organizational practices of *systems management.*[3]

The computer industry developed largely (although not completely) from the office machine industry. Each of the four major companies, International Business Machines (IBM), National Cash Register (NCR), Burroughs, and Remington Rand, initially developed single machines such as tabulators, cash registers, and adding machines. Over time these companies expanded their product lines and developed "systems" of machines, such as tabulators connected with sorters and printers. All of the firms were fiercely competitive and thrived on product innovation. The shift to electronic digital computers came after World War II, in each case with significant funding and direction from the U.S. military. For the computing industry, the systems approach inherited from office machines manufacturing mixed with systems approaches developed in the military to yield a powerful formula for innovation.[4]

The aerospace industry derived from a similar mix: from the pre-WWII aircraft industry and post-WWII ballistic missile developments. Like office machines that eventually evolved to include electronics, aircraft began as mechanical devices and grew progressively more complex, ultimately including mechanical, hydraulic, and electronic technologies. By the end of World War II, Air Force bombers, such as the B–29, and jet fighters were complex systems that required significant up-front planning for their design and manufacture. Ballistic missiles complicated matters further by requiring total automation and rocket technologies along with other components typical of aircraft. In the aircraft and early space industries (soon to be known as "aerospace"), one either welcomed rapid innovation or went out of business because the military demanded state-of-the-art performance for its aircraft, missiles, and space systems.

Together, the military, academia, and private industry developed organizational processes to accommodate rapid technological change.[5] Although ultimately successful in driving necessary change, these close relationships and new processes formed what President Dwight D. Eisenhower called the "military–industrial complex." Warning that "the government contract becomes virtually a substitute for intellectual curiosity," Eisenhower believed that the resulting intimate relationship between the public and private sectors formed a new phenomenon in U.S. society—a "technocratic" state driven not by democratic principles but by the dictates of efficiency and military competition. Systems management is at the heart of such relationships. Although some worried that these conjunctions corrupted American society, the creators of systems management intended to protect society by tying together organizations typically held at arm's length: the military, private industry, and academia. They did so in the name of technical progress through research and development.[6]

Successful research and development (R&D) requires several elements: innovative research, focused development, and the means to move an idea or technology from one endeavor to the other. For R&D to be successful, the gap between development and operations also must be bridged.

Excellent research generally has the following common qualities: First, the researchers are of high caliber and have freedom to pursue their own interests, determining the goals of the research and its products. Second, funding and the facilities necessary to perform the research are available. Last, researchers generally organize in an ad hoc manner that they establish for themselves.

Development, however, requires an entirely different context. Whereas research is often open-ended, development requires a specific goal to which the entire team adheres. Whereas research management is generally rather decentralized, a successful development project often features a strong manager who maintains a firm grip on the project. Development usually consumes far greater amounts of money than does research, thus making errors more costly and planning much more important. Because the end result of the development is a product that is used in practical application, developers must account for how the product will be operated by its end users and design the product accordingly.

Making the jump from research to development often is the trickiest part of R&D. Decision makers must select which research projects should move forward into much more costly development. Research managers sometimes promote their research products as solutions to existing, real-world problems; in other cases, developers or end users search for solutions to known problems.

The USAF and the Culture of Innovation

The aerospace and computing industries developed organizational methods that spanned the spectrum from free-wheeling research to rigorous development. They did so primarily through their interactions with the military and, most important, with the Air Force. By the end of World War II, military officers recognized the importance of such novel technologies as the atomic bomb and radar, and created new organizations to continue the productive alliance among the armed services, scientists, and technologists. When the threat of Soviet nuclear capabilities emerged, the services again actively recruited the technical experts to help them create technical responses—ballistic missiles and air defense systems. Ultimately, those two technical developments challenged the technologists and their military patrons, and led to significant new practices for R&D.

Ballistic Missiles

Ballistic missiles developed from simple rocketry experiments between World Wars I and II. Experimenters such as Robert Goddard and Frank Malina in the United States, Wernher von Braun in Germany, Robert Esnault-Pelterie in France, and Valentin Glushko in the Soviet Union found that rocketry was a dangerous business. All of them had their share of spectacular mishaps and explosions before achieving occasional success.[7] Why was rocketry so difficult to achieve in practice when it was so simple in theory?

The most obvious reason for the difficulty of rocketry is the extreme volatility of the fluid or solid propellants. Aside from the dangers of handling exotic and explosive materials such as liquid oxygen and hydrogen, alcohol, and kerosene, their combustion must be both powerful and controlled. This means that engineers must channel the explosive power so that the heat and force neither burst nor melt the combustion chamber or nozzle. Rocket engineers learned to cool the walls of the combustion chamber and nozzle by maintaining a fuel flow near the chamber and nozzle walls to carry off the excess heat. They also enforced strict cleanliness in manufacturing, because impurities or particles could and did lodge in valves and pumps, with catastrophic results.[8] Engineers controlled the explosive force of combustion through carefully designed liquid feed systems that smoothly delivered fuel. Instabilities in the fuel flow could and did cause irregularities in the combustion, which often careened out of control, leading to explosions.

Hydrodynamic instability also could ensue if the geometry of the combustion chamber or nozzle was inappropriate. Engineers learned through experimentation the proper sizes, shapes, and relationships of the nozzle throat, nozzle taper, and combustion chamber geometry. Because of the nonlinearity of hydrodynamic interactions, experimentation rather than theory determined both the problems and the solutions. For solid fuels, the shape of the solid determined the shape of the combustion chamber. Years of experimentation at the Jet Propulsion

Laboratory (JPL) eventually led to a star configuration for solid fuels that provided steady fuel combustion and a clear path for the exiting hot gases.

After engineers solved those many difficulties, they faced the challenging problems of aerodynamic stability. Aircraft typically are optimized for a limited range of atmospheric pressures and altitudes. Rockets, however, travel through the atmosphere from the surface of the earth to the vacuum of space at speeds that range from zero miles per hour to far greater than the sound barrier. In flight they encounter the full range of atmospheric pressures, and must be designed to fly through all of them. They fly through fierce crosswinds as well as a complete vacuum.

As a rocket flies, its center of gravity changes continuously.* When it is fully fueled a rocket's center of gravity is determined primarily by the location of its fuel. It is not much of an exaggeration to describe a rocket as a flying fuel tank. As its fuel burns and the tanks empty, the center of gravity shifts dramatically because the mass distribution of the rocket then is determined by the remaining hardware. The loss of fuel also changes the rocket's resonant frequencies, the frequencies at which the structure bends most readily. Staging compounds this problem because the rocket sheds its lower stages as it flies through the atmosphere. Structural engineers have characterized the flight of the Titan III rocket as similar to balancing a wet noodle while pushing it from behind through the atmosphere.[9]

Rocket engines create severe structural vibrations. Aircraft designers recognized that propellers caused severe vibrations, but only at specific frequencies related to the rotation rate of the propellers. Jet engines posed similar problems, but at higher frequencies corresponding to the more rapid rotation of turbojet rotors. Rocket engines were much more problematic because their vibrations were not only large, but occurred at nearly random frequencies over a very wide range. This caused breakage of structural joints and the mechanical connections of electrical equipment, thus making it difficult to fly sensitive electrical equipment such as vacuum tubes, radio receivers, and guidance systems. Along with the vibration of the engines were vibrations caused by fuel sloshing around in the emptying tanks and fuel lines. These so-called pogo problems could be tested only in flight.

Vibration problems generally could not be solved through isolated technical fixes. Because vibration affected electrical equipment and mechanical connections throughout the entire vehicle, this problem often became one of the first issues that brought together the structural engineer, the propulsion expert, and the electrical engineer. In the 1950s vibration problems led to the development of the new discipline of reliability and to the enhancement of quality assurance, both of which crossed traditional engineering disciplines.[10]

---

*The center of gravity is the location at which an object can be perfectly balanced.

The USAF and the Culture of Innovation

Rocket engineers tried to reduce engine and pogo vibrations as much as possible but they could not eliminate them. Engineers in other disciplines had to protect their equipment as best they could. First, engineers placed stronger emphasis on the selection and testing of electronic components. Parts to be used in missiles had to pass stringent tests, quality checks, and inspections. Second, technicians assembled and fastened electronic and mechanical components to electronic boards and other components using carefully crafted techniques of soldering and fastening. They could use only certain kinds of soldering methods or fasteners that passed the military's new vibration standards. Third, engineers subjected assembled components and the entire vehicle if possible to vibration tests. This involved the development of large vibration or "shake" tables. Fourth, the military required that trained quality assurance personnel witness and document all of these new procedures. Military authorities gave them independent reporting and communication channels to avoid possible pressures from contractors or government officials. The new methods involved novel ways of organizing and standardizing the use and assembly of existing technologies, and were among the first methods given a "systems" label. They became standard features of systems management.[11]

Only when engineers solved the vibration problem could they be certain the rocket's electronic equipment would send the signals necessary to determine its performance. Unlike aircraft, rockets were automated. Although automatic machinery had grown in importance since the eighteenth century, rockets took automation to another level because in flight they could not be tended by hand. Aircraft could be piloted because the dynamics of an aircraft moving through the air were slow enough that pilots could react quickly enough to correct deviations from the desired path and orientation of the aircraft. The same does not hold true for rockets. Combustion instabilities inside the rocket engines occur in tens of milliseconds and explosions happen within one hundred to five hundred milliseconds, leaving no time for a pilot to react.

Because rockets were fully automated and because they went on a one-way trip, never to return, determining if a rocket worked correctly was (and is) problematic. Engineers developed sophisticated signaling equipment to send performance data to the ground. Assuming that this telemetry equipment survived the launch and vibration of the rocket, it sent sensor data to a ground receiving station that recorded them for later analysis. Collecting and processing these data were some of the first aerospace applications of analog and digital computing technology. Data collection had to be preplanned before each test flight and coordinated across all rocket components to ensure that each engineer received a share of limited telemetry data. Engineers used the data to determine if their subsystems worked correctly or, more important, to determine what went wrong if the subsystem did not function.[12]

Extensive use of radio signals caused more problems. Engineers used the signals to send telemetry to ground stations and to send guidance and destruct signals from the ground station to the rocket. They had to design the electronics and wiring carefully so that electromagnetic waves from one wire did not interfere with other wires or radio signals. As engineers integrated numerous electronic packages, the interference of these signals occasionally caused failures. The analysis of "electromagnetic interference" became another systems specialty.[13]

Automation also included the advanced planning and programming of rocket operations known as "sequencing." Rocket and satellite engineers developed automatic electrical or mechanical means to open and close propulsion valves, fire pyrotechnics to separate stages, release the vehicle from the ground equipment, and otherwise change rocket functions. These sequencers usually were specially designed mechanical or electromechanical devices, but soon the capabilities of digital computers were used in their place. A surprising number of rocket and satellite failures resulted from improper sequencing or sequencer failures. For example, rocket stage separation required precise synchronization of the electrical signals that fired the pyrotechnic charges with the signals that governed the fuel valves and pumps controlling propellant flow to the upper-stage engines. Because engineers sometimes used engine turbopumps to generate electrical power, failure to synchronize the signals for separation and engine firing could lead to a loss of sequencer electrical power. This in turn could lead to collision between the lower and upper stages, to engine explosion or failure to ignite, or to no separation.[14]

Many of the technical problems of rocket design led engineers to develop solutions that emphasized rigorous processes and procedures. Problems of rocket vibration, component reliability, and particle contamination all lent themselves to process solutions: rigorous control of cleanliness, part selection and testing, and soldering and coating techniques in manufacturing. Other problems, such as electromagnetic interference, telemetry data gathering, and sequencing, required system analyses combined with system testing to ensure proper functioning of the integrated components.

Although rocketry remains a risky business, those new organizational processes made rockets far more reliable than when ballistic missiles first flew. Failure rates of early missiles ranged from 40 to 60 percent, but current-day launch vehicles derived from those missiles typically succeed from 90 to 95 percent of the time. Rocketry shared reliability problems with the Air Force's defensive response to nuclear weapons—air defense and early warning systems—later known as "command and control" systems.[15]

Automation of Command and Control

Defense against attack from bombers or missiles essentially is a problem of detection, interpretation, and response. First, there must be some means to detect whether aircraft or missiles are approaching the territory to be defended. All sensor data regarding the approaching objects must be relayed to some centralized location(s) where military officers determine if the objects are hostile and, if so, what response should be made. Once a response decision has been made, commands must be relayed to the appropriate organizations to attempt intercept and destruction of the incoming objects, and to other organizations to take such actions as moving to shelter, launching a counterstrike, and so on.

Early defense systems consisted primarily of human beings who collected, transmitted, analyzed, and responded to the information. By the eve of World War II, radar significantly improved the capability to detect aircraft, and telegraph and telephone lines generally transmitted the signals to centralized command posts where officers determined which fighter units should intercept the incoming raids. These means sufficed throughout that war until the German A–4 (V–2) ballistic missile appeared in Allied skies. This weapon was particularly disturbing because it arrived with virtually no warning and descended from space so rapidly that no interception was possible. Despite those frightening capabilities, ballistic missiles carried only conventional explosives and were so inaccurate as to be useless as military weaponry.

Three new technologies rendered those old-style air defense systems obsolete: nuclear weapons, jet engines, and ballistic missiles. While nuclear weapons made the destructive power of a single bomber or missile catastrophic, jet engines and ballistic missiles significantly decreased the reaction time available to the defenders. Those two factors made it essential to detect approaching objects, assimilate and interpret the data, and respond all very rapidly. Decreasing reaction time made automation of the entire process an increasingly attractive option. This in turn made the electronic digital computer a critical military technology.

One of the most important problems of early computer systems was the reliability of the computer hardware. Because computers were (and are) made from thousands of electronic components, and most or all of these had to function properly for calculations to proceed correctly, the failure rate of any given component had to be very low. Electronic component reliability was an important concern because many electronic components of the time, such as vacuum tubes, had failure rates that were quite large compared with the rates required for successful computer operation. For air defense applications in which the computer system had to run continuously, the problem took on critical proportions.

Most computer builders took the problem seriously, but Jay Forrester's group at Massachusetts Institute of Technology (MIT) made reliability an obsession mainly because of the air defense application of its Whirlwind computer. The Whirlwind group calculated the reliability it would need from its computer and concluded that extraordinary measures would be needed to achieve it. For one component, the computer memory, the measures involved inventing a new technology that became the industry standard for the late 1950s and 1960s—the magnetic core memory. For the most part, however, solving the reliability problem did not require engineering genius so much as it required strict attention to manufacturing, assembly, and testing of components and the system integrated from them. Forrester's engineers recognized that most parts failed because they were stressed in some way, so they insisted on components that could withstand severe stresses and then operated them so that the components never came anywhere near their limits. Assembly and testing methods had to be designed and followed rigorously, and manufacturers were monitored strictly to ensure that they followed those procedures. In short, reliability came through strict attention to processes that was achieved by social control, not through some magical new artifact.[16]

The social aspects of engineering also took priority of place in the development of software. Software is a pure process, nothing more than step-by-step procedures encoded as machine instructions. Each software routine reflects the thought process of the programmer, and the overall program is an assemblage of routines and, hence, of the thoughts of its programmers. Unclear or confused thinking yields confused software, and a chaotic organization yields equally chaotic software.

Small software programs are relatively easy to assemble and test because members of a small team of programmers easily can communicate the relevant information about their programs to the other team members. Large software programs, however, require correspondingly larger teams. Beyond a certain size, individual team members no longer can keep track of all information from all other members. These communication problems result in software routines unable to communicate properly with each other and that leads to various malfunctions. The solution to these problems is to organize the programmers and their communication efficiently, and to reflect this enhanced organization through the design of the software. For example, if you organize the programmers such that each task is largely self-contained, then the corresponding software routines will need less communication with other routines and the number of possible communication (and hence coding) errors will be reduced. If all programmers need a set of information, then separating that information into routines accessible by all other routines simplifies the communication process for the programmers and their routines.[17]

When software is the tool to organize the communication of other organizations, as is the case for command and control (C^2) systems, the problem is confounded even further. Command and control is inherently a social process. As described earlier, an air defense system must collect information, assemble it into a form useful for analysis and decision making, make appropriate decisions, issue commands, and verify that the commands have been followed. This depends on top-level policy decisions about the importance of air defense, on the overall defense strategy, and on the intimate details of air tactics. Any of those can change based on external changes in the nature of the threat; political changes and agreements at home or abroad; changes in offensive, defensive, or communications technologies; or even on the biases of individual commanders.

The USAF and the Culture of Innovation

Before computers existed, command and control systems consisted of human beings who communicated with each other either directly or by telephone, written correspondence, encryption techniques, and so on. Whatever other faults they might have, humans are quite flexible in adapting to changing circumstances. If a commanding officer reorganizes the people under him or her, or changes their operating procedures, those people can accommodate the changes with relative ease. The same is not true of computers or their software.

To the extent that computing systems and software are elements of a command and control system, they must readily be reprogrammable to meet changing circumstances. Another alternative is to automate only those few portions of the system in which changing circumstances do not affect the procedures that must be carried out. Developers must therefore be aware of the organization and processes of their own development groups and those of the organizations that the C^2 system is intended to assist. To take a simple example, if initially the C^2 system issues commands to four organizations, and then a new organization becomes involved, new programs must be written to issue commands to the new organization. Similarly, if the processes change so that different instructions go to an organization, then old programs must be changed to reflect the new instructions. To an extent quite unlike ballistic missiles, C^2 systems undergo continuous change. They are a product forever "in work" and never completed.[18]

If software reflects the detailed processes of those organizations with which it works, then those who write the requirements for the software must master the knowledge of these organizations to be able to reduce it to computer instructions. The same holds true for the technologies with which the C^2 system connects. For both C^2 systems and ballistic missiles, those who integrate the system must master to a significant degree the knowledge of the diverse technologies of which the system is made. This problem of technical complexity was one of the crucial new issues that the system developers of the 1950s had to face.

For both ballistic missiles and command and control, technical complexity led to the development of new organizational processes created by those who were most involved. The approaches of scientists, engineers, and managers differed because of their diverse backgrounds, but they all contributed to creating the Air Force's new weapons and technologies.

Technical Complexity and Systems Approaches

Beyond the particular problems of ballistic missiles and command and control systems, both developments shared the fundamental problem of complexity. Three factors made them more difficult than prior technologies. First, their novelty made it difficult for engineers in the aircraft and office machine industries to assimilate them into products. Prior to World War II, structural and mechanical expertise defined the skilled workers in those industries. After the war, that expertise no longer sufficed as electronics, rocket and jet propulsion, and guidance and control technologies became critical. Second, the depth of knowledge required in the new disciplines was such that it could not be mastered easily. For the mechanically adept engineers of the earlier period, the mathematical and analytical skills necessary in the new disciplines were daunting. Finally, the new technologies were heterogeneous. Each of the new disciplines required radically different kinds of skills than did the earlier mechanical aptitudes. Chemical processes critical for rocket propulsion differed from the electronic skills necessary for computer development, which differed a great deal from the analytical methods and precision machinery required for guidance and control systems. All of these factors made it impossible for any one person to master in depth all of the skills needed to build a ballistic missile or an air defense system.

These complexities were reflected in increasing cost and time to develop the new weapons. According to Ellis Johnson, the head of the Operations Research Office at Johns Hopkins University,[19]

> The effect of increasing physical knowledge on the cost of weapons in a weapons system has been very great in terms of money and complexity. . . . It can be seen that this cost has increased ten-fold from 1945 to 1955. . . . In aircraft gas turbines the number of parts has increased from 9,000 in 1946 to 20,000 in 1957. Of precious engineering hours, 17,000 were required to produce a fighter aircraft in 1940, and 1,400,000 in 1955. . . .[20]

In this new, perplexing environment, the "jack-of-all-trades" technical generalist assumed critical importance. He (in the 1950s this person was invariably male) had several names. Generalist engineers became known as "systems engineers." Scientists called their generalists "operations researchers." Those who managed the technical projects, regardless of their background, became "project managers." Organizations that wanted to develop the new technologies such as ballistic missiles and command and control systems placed these new generalists to bridge the gap between the administrators and the technical experts. The generalists in turn worked with technical specialists to incorporate the radical new technologies.

Because new technologies such as nuclear weapons, radar, computers, and rocket propulsion had only recently been developed, military and industrial organizations had little choice but to include the physicists and "rocket scientists" (usually engineers) in their design teams. That explains the sudden interest in multidisciplinary teams during and immediately following the Second World War.

The Manhattan Engineer District was extremely influential as an early example of a project-based multidisciplinary team of scientists and engineers. It is not too surprising that this large project was given to a leading manager from the Army Corps of Engineers, an organization well known for managing large technical projects. Gen. Leslie Groves administered the project with a three-person staff and made major decisions with a small committee that consisted of himself, influential scientists Vannevar Bush and James Conant, and representatives of each of the services. Army officers directed day-to-day operations at each field site, most of which had traditional hierarchical organizations augmented by secrecy. Because of technical and scientific uncertainties, the project developed two bomb designs and three approaches to create the fissile material.

At Los Alamos in 1942–43, Director Robert Oppenheimer wrested a degree of freedom of speech for the scientists and ensured that they remained civilians. To maintain open communication and in deference to academic norms, Oppenheimer kept the initial organization as nonhierarchical as possible, adopting the loose department structures typical of universities. That changed by the spring of 1944 when the project accelerated work on the complex implosion design. As research and development teams grew, the project needed and obtained stronger managers like Robert Bacher and George Kistiakowsky, who transformed the organization from an academic model to divisions organized around the end product—a project organization.[21] Along with the German V–2 project, made known to U.S. military and industrial leaders through Theodore von Kármán's report on German scientific and technical innovations at the end of World War II, the Manhattan Project became the prototypical model for military project management of complex technology. For example, in 1953 when Brig. Gen. Bernard Schriever considered how he should manage the Air Force's intercontinental ballistic missile (ICBM) program, he thought about the Manhattan Project as a model. Although he ultimately developed his own methods, the bomb project was his initial guide.[22]

Engineers developed their own approach to technical complexity: systems engineering. Systems engineering (SE) has at least three separate roots. The office machine industry developed a version of systems engineering in the 1920s and 1930s with the now-familiar problem of uniting a central, generalized machine like a programmable tabulator or mechanical calculator with peripherals such as card sorters, punches, readers, and printers.[23] Another version of systems engineering was formulated at American Telephone and Telegraph (AT&T), where engineers maintained and upgraded a complex system of telephony. Those engineers began to call their work "system engineering" because of the constraints the existing technology imposed on their effort. Because of the organizational separation between Bell Telephone Laboratories on the one hand and Western Electric (AT&T's manufacturing arm) on the other, AT&T personnel developed formal documentation methods to ensure clear communication between the laboratory scientists and engineers in the labs and the manufacturing engineers at Western Electric.[24] A third version of systems engineering grew from the development of military technologies at MIT Radiation Laboratory. That version was particularly influential because its creator, physicist Ivan Getting, later interacted with important military and civilian personnel.

Gen. Bernard A. Schriever is considered the father of the U.S. Air Force intercontinental ballistic missile program. Here he poses with models of the many systems for which he was responsible.

Getting was in charge of the development of the Navy's Mark 56 Fire Control System. Recognizing that the radar and gun directors working together behaved differently from the way one operated separately because of noise, he noted that "the specifications on each unit should be written with full consideration of the features and capabilities of the other."[25] This led to the creation of a new role, the *system integrator* who was neither the manufacturer nor the user. This person would make technical information available to government and contractor organizations; check and critique designs; send representatives to participate in conferences; report on project progress; participate and establish procedures for prototype, preproduction, and acceptance testing; and assist in training programs. Getting arranged for the Radiation Laboratory to receive relevant correspondence, drawings and specifications, and notification of significant tests and conferences, and to have access to contractors, engineers, and equipment.[26] Getting's idea of systems integration spread through his influential positions, including his membership on the Air Force's Scientific Advisory Board (SAB), his technical directorship for Air Defense Command (ADC), and his presidency of the Aerospace Corporation.[27]

The USAF and the Culture of Innovation

Radar development was also influential in creating a third systems approach: operations research (OR).[28] Just prior to and in the early years of World War II, British scientists greatly improved the efficiency of the air defense system, a major factor in the Royal Air Force's victory over the German *Luftwaffe* in the Battle of Britain. They did so through mathematical analysis, application of physical laws, and application of these ideas to the interactions of radar operations with the air defense system. Their success was such that in August 1940 the military asked the scientists to improve antiaircraft gun ranging and targeting. Physicist P. M. S. Blackett's "circus" of scientists promptly increased the efficiency of the antiaircraft system, and then went on to improve airborne radar ship and submarine detection. After these initial successes, the British military gave operational researchers substantial latitude to communicate at all levels of the military hierarchy and to pick their own problems, which included convoy size and tactics, antisubmarine air tactics and submarine detection, and bomber formation size and tactics.

Operations research quickly spread to the United States, leading to similar successful applications in the Navy and the Army Air Forces (AAF). After the war, the military established permanent operations research organizations, including the Navy's Operations Evaluation Group, the Army's Operations Research Office at Johns Hopkins University, the Army Air Forces Operations Analysis Groups in each command, and the RAND Corporation to provide long-term research. Secretary of Defense James V. Forrestal established the Weapons Systems Evaluation Group in 1947 to serve the Joint Chiefs of Staff by providing "rigorous, unprejudiced and independent analysis and evaluations" of present and future weapons.[29]

Operations research was influential at the nonprofit "think tank," the RAND Corporation. RAND researchers extended the techniques to investigate the potential value of future systems. They called this future-oriented operations research "systems analysis." After some unsuccessful attempts at comprehensive analysis of air warfare, they focused on smaller problems for which systems analysis proved successful.[30]

Like operations research, systems analysis used teams of mathematicians, scientists, engineers, managers, economists, and military officers.[31] Systems analysts borrowed and modified mathematical and physical methods when necessary and applied them to complex human–machine systems. Typical methods included game theory, probability, and applications of physical laws such as classical mechanics and electromagnetic theory for radar. Extensive use of computing techniques led operations researchers in the 1950s to develop new mathematical methods like linear programming as computational aids.[32]

For both operations researchers and systems engineers, the interrelationships of large weapons systems with external, social factors were intimate and explicit. The new systems had to take into account existing systems and organizations. As Simon Ramo put it, "The economic, military, governmental, and even sociologic [*sic*] considerations enter more and more frequently into systems

Physicist Ivan A. Getting created an influential version of systems engineering while developing military technologies at MIT's Radiation Laboratory. He later became Schriever's boss as Assistant for Evaluation and was the first president of the Aerospace Corporation.

engineering decisions. . . ."[33] They also had to account for the "possible interactions among men and machines" within existing organizations. Scientific and technological achievements had, in Ellis Johnson's opinion, made coordination increasingly difficult. Increasing complexity "made it impossible to start with a fixed plan of action" because all of the technical and social elements interacted so tightly with each other.[34]

What were the new factors that complicated the lives of scientists and engineers? For scientists who became operations researchers, human beings were the new factor. Improving the efficiency of the military meant improving the efficiency of humans and machines together. Operations research modeled humans and their actions in the same mathematical ways that physicists and mathematicians used for their normal research. By contrast, engineers were familiar with the operation of complex human–machine systems but they faced the arrival of new science-based technologies. By the late 1930s and during WWII, electronics became crucial, starting with guidance, communication, and radar equipment. In the late 1940s and early 1950s, nuclear warheads, and rocket propulsion and automatic control systems made their appearance with unmanned guided missiles and rockets, thus adding more unfamiliar disciplines to the mix. The appearance of computers and their programming contributed yet another radical technology. Systems engineering was the way scientists and engineers integrated the new specialties into the familiar tasks of design and manufacturing.

Both operations researchers and systems engineers agreed that they performed services useful to management. In 1954 Ellis Johnson claimed, "with each passing day, it [operations research] is increasing its capability of helping management to solve complex action problems and make major decisions."[35] Arthur Hall, a systems engineer from Bell Laboratories, stated that systems engineering provided "management with as much information as possible needed to guide and control the overall development program."[36] Both promoted their disciplines as the technical arm of the managerial technocrat.

Despite the lack of new theoretical content in operations research or systems engineering, a few schools with practical interests began to teach these topics. In 1948 MIT established with the Navy a course in the nonmilitary applications of operations research. The University College in London, Case Institute of Technology, Columbia University, and Johns Hopkins University established operations research courses by the mid-1950s.[37] In 1952 MIT developed a weapons systems engineering course associated with Charles Stark Draper's Instrumentation Laboratory, and 118 Air Force officers had graduated from the course by 1958.[38] G. W. Gilman of Bell Laboratories began informally teaching systems engineering at MIT in 1950, and in December 1954 he started a systems engineering course as part of Bell's Communications Development Training Program. Recognizing the similarities between operations research and systems engineering, by 1962 the University of California at Los Angeles, the University of Pennsylvania, the University of Michigan, and Johns Hopkins University offered graduate courses titled "Operations Research and Systems Engineering."[39] Textbooks and journals eventually followed. As operations research and systems engineering diffused through books, journals, conferences, and the movement of technical personnel, the aerospace and computing industries soon came to view these methods as the standards for coordination and analysis of typical projects.[40]

Systems Management

"Systems management" is the name given to the combination of organizational methods developed in the United States Air Force in the 1950s and early 1960s. Combining ideas from project management, systems engineering, and operations research, it became the standard for large-scale project development by 1960, and later the standard for the Department of Defense (DOD). Management writers typically present systems management by showing four development phases and then describing the techniques that should be applied in each phase. Figure 1 presents those phases and their requisite techniques.[41]

Figure 1

Standard Presentation of Systems Management

Phase A Systems analysis System trades	***Conceptual Studies*** Award multiple, small fixed-price contracts
Phase B Systems engineering Subsystem trades	***Preliminary Design*** Award two to three medium fixed-price contracts
Phase C Systems engineering Subsystem engineering	***Detail Design*** Award one large cost-plus-incentive-fee contract Establish project control room Perform preliminary design review Establish change control over interface Perform critical design review Establish change control over subsystems
Phase D Subsystem manufacturing System integration	***Manufacturing and Testing*** Perform subsystem test Establish change control over procedures Perform system interface (static) test Perform system (dynamic) test Perform first article configuration inspection Perform flight readiness review Perform flight test (if applicable)
	Operations

For a systems management view, projects start as ideas for scientific or commercial experimentation. Scientists and engineers generate ideas for experiments and missions, and perform trade studies to determine which are feasible and most worthy of study. After selection of a mission, the military performs or contracts for detailed design studies that determine the cost and schedule to construct and operate the proposed vehicle. The specifications and estimates from these studies form the basis for a final design competition. After contractor selection, the military and the contractor progressively develop design specifications and a design, which they assess in formal design reviews. The design review teams recommend changes that the contractor incorporates into the design, and then manufacturing begins. During manufacturing, strict standards enforce high-quality manufacturing practices and component selection. Once built, the contractor assembles and tests the system in environments that simulate the environment in which it will operate. These tests reveal remaining design flaws that the contractor corrects. Presented in this way, systems management appears as a highly logical, self-evident, and self-contained process.

The pervasiveness and apparent logic of these methods at the start of the twenty-first century make it difficult for military and civilian managers to understand that systems management did not develop as a unit. In fact, it is a melange of techniques developed in response to distinct problems of the 1950s and 1960s. By revisiting the history of their development, we profitably may rediscover where and why these methods came into existence. By doing so we have a much better chance of selecting which of those elements of systems management are relevant to the specific circumstances of projects and organizations in the twenty-first century, and discarding or replacing the rest.

To elucidate systems management, I will divide it rather differently from the standard presentation shown in management texts. Instead, Table 1 shows the social groups that promoted specific techniques of systems management. When we view it in this manner, we begin to understand how systems management represents the values and concerns of its contributing social groups. We then will be in a better position to understand how and why it fits the aerospace and computing industries reasonably well, and why it does not operate so well outside of those contexts.

Systems management developed and spread through the influence of four career groups: military officers, scientists, engineers, and managers. Each promoted those aspects of systems management that were congenial to their objectives, and fought those that were not. Because each group had different concerns and goals, their favored systems management conceptions and processes also differed.

For example, the military's conception of "concurrency," which promoted speed of development, ran counter in a number of ways to the managerial idea of "phased planning," which promoted cost control. Scientific conceptions of "systems analysis" differed from the engineering idea of "systems engineering." Informal working groups favored by scientists and engineers conflicted with hierarchical structures and processes typical of industry and the military. Occasionally the conflicts found their way into print, but more often than not the disputants resolved them internally through bureaucratic infighting. The winners of these battles usually imposed new structures or processes that promoted their conceptions and their power within and across organizations.

Table 1

Social Groups Involved in Systems Management
and the Techniques They Favored

Military Officers	Scientists	Engineers	Managers
Concurrency	Operations research	Systems engineering	Phased planning
Technical direction	Systems analysis	Systems integration	PERT/CPM
Control rooms	Cost–benefit analysis	Contractor penetration	Contractor penetration
Project organization	Functional organization with committees	Functional organization with committees	Project and matrix organizations
Configuration management		Design freeze/ change control	Configuration management
Multiple source competition	Quantitative reliability	Qualitative reliability	Work package management
		Quality assurance	Quality assurance
System testing		Environmental and system testing	

CPM = Critical Path Management; PERT = Program Evaluation and Review Technique.

A Social Story of Air Force R&D

During World War II and in the early Cold War in the late 1940s and 1950s, scientific prestige reached a height that probably has never been equaled. The development of the atomic bomb, radar, and rockets during the war presaged the hydrogen bomb and guided missiles that would come after it. Military leaders harnessed scientific expertise through their lavish support of scientists and the development of new laboratories and research institutions. Scientists in turn provided the military with technical and political support to develop new weapons.[42] The alliance of these two groups led to the dominance of the policy of "concurrency," and to the science-dominated military research policy of the 1950s.[43]

In military parlance, concurrency means conducting research and development in parallel with the manufacturing and production of a weapon. More generally, it also can be used to describe any parallel processes or approaches. Concurrency met the needs of military officers because they perceived critical external threats that they believed demanded an immediate response. Put differently, for military officers to acquire significant power in a civilian society, the society must believe in a credible threat that must be countered with military force. Under such conditions, which clearly prevailed in the early Cold War, military leaders rapidly developed countermeasures. For the armed forces, external threats and rapid technological development went hand-in-hand.

Concurrency also met the needs of scientists who purportedly held the keys to novel "wonder weapons." Radar and nuclear weapons developed from scientific research and thus confirmed the stature of postwar science. Even when scientists had little to do with a major technological advance, as in the cases of jet and rocket propulsion, society often deemed the engineers involved to be "rocket scientists." Scientists did little to discourage this misperception. Scientists had a natural tendency to predict and foster novelty because discovery of new natural laws and behaviors was their business. Novelty required scientific expertise, whereas "mundane" developments could be left to engineers. Research policy emphasized the funding of physical scientists and research engineers, with few questions asked. It was believed that scientists knew best what avenues of research would lead to profitable application and they should be free to choose the methods and directions of research. The military's role was to distribute funding and consult with the scientists about the new "toys" that inevitably would follow.

By 1959, however, Congress began to question more severely the military's methods, particularly because their weapons cost far more than predicted and they did not seem to work.[44] A number of embarrassing rocket explosions and air defense system failures spurred critical scrutiny. Although Sputnik and the Cuban Missile Crisis dampened criticism somewhat, military officers had a difficult time explaining the apparent ineffectiveness of its new wonder weapons and systems. Missiles that failed about half the time were efficient deterrents neither to Soviet aggression nor to congressional investigation. It seemed foolhardy for the president to order a nuclear counterstrike based on an early warning system that mistook the rising moon as a Soviet missile attack. Military leaders needed to prove they could make these new technologies work and do it within a predictable budget. Managers and engineers were well suited for such problems.

There are two types of engineers: researchers and designers. Engineering researchers for our purposes can be considered very similar to scientists, except that their quest involves technological rather than "natural" novelty. They work in academia, government, and industrial laboratories and have norms involving the publication of scientific papers, development of new technologies and processes, and the diffusion of knowledge. By contrast, engineering designers spend most of their time designing, building, and testing artifacts.† Depending on the product, their success criteria involve cost, reliability, and performance factors. Design engineers have little time for publication, and proof of their claim to expertise is a successful product.

Even more than design engineers, managers must pay explicit heed to cost considerations. Their claim to expertise lies in the effective use of monetary resources to accomplish organizational objectives. Managers measure their power from the size and funding of their organizations so they have conflicting desires to use their resources efficiently, which decreases the size of the organization, and to make their organizations grow so as to acquire more power. Ideally, a manager achieves objectives in the most efficient way and acquires more power by acquiring other organizations or tasks. In a commercial organization, growth occurs through the development of a successful product, which then requires the development of mass-production techniques. This avenue for growth does not typically exist in government organizations, however, and more often occurs through expansion of tasks or expansion of the organization through less efficient use of resources. Managers, like engineers, lose credibility if their end products fail.[45]

As ballistic missiles and air defense systems went through their "teething period" in the late 1950s, military officers and aerospace industry leaders had to heed congressional calls for greater reliability and cost control. As a consequence, managerial and engineering design considerations came to have more weight in technology development. Managers responded by applying more rigorous cost-accounting practices, and engineers responded with more testing and analysis. The result was not a "low-cost" design but a more reliable product whose cost was high but predictable. Engineers gained credibility through the success of the missiles, and managers gained credibility through successful prediction of the cost. Because of the high priority and visibility of space programs, congressional leaders in the 1960s did not mind high cost, but they would not tolerate unpredictable costs or spectacular failures.

Systems management consisted of these informal processes eventually encapsulated in written procedures. Scientists and engineers tried to maintain informal methods that kept communication and information to themselves and out of the hands of managers and the military. Military officers occasionally tried to keep methods informal for the same reasons, as when Schriever's new missile organization initially used informal methods to prevent others from interfering (a matter that will be discussed in Chapter 2). Military officers, like managers,

†An "artifact" is any useful tool or device.

however, usually created and imposed standardized procedures to maintain their control. Their efforts resulted in the development of new methods to organize research and development.

Conclusion

The methods of systems management encompass the analytical methods of scientists, the design and testing methods of engineers, and the surveillance and organizational methods of managers and military officers. Those groups used them to foster their own conceptions and authority, and each applied methods with which they were familiar to the new problems they faced. Scientists applied mathematical techniques to analyze new and existing technologies. Engineers created design and testing methods to develop new technical systems. Managers and military officers applied and modified new techniques to control costs, schedules, and performance.

Systems management was the result of these often-conflicting interests and objectives. It was (and is) not a monolithic entity but rather a mix of techniques combining the interests of each contributing group. We can define *systems management* as a set of organizational structures and processes whose goal is rapidly to produce a novel but dependable technological artifact within a predictable budget. In that definition we can see each of four groups: Military officers demanded rapid progress. Scientists desired a novel product. Engineers wanted a dependable product. Managers sought a predictable cost. Those four social groups created new organizational structures and processes for R&D, and in so doing laid the foundations for the future success of the U.S. aerospace and computing industries.

Notes

1. Col Norair M. Lulejian, Dep for Tech, SSD, "Scheduling Invention," paper presented to AFSC Management Conference, Monterey, Calif, May 2–4, 1962, AFHRA, microfilm 26254, p 1-4-1.

2. Ross Thomson, "The Firm and Technological Change: From Managerial Capitalism to Nineteenth-Century Innovation and Back Again," *Business and Economic History* 22 (Winter 1993), pp 99–134; Bruce Kogut, ed, *Country Competitiveness: Technology and the Organizing of Work* (Oxford: Oxford University Press, 1993); Michael Porter, *Competitive Advantage of Nations* (New York: The Free Press, 1990); Nathan Rosenberg, *Exploring the Black Box: Technology, Economics, and History* (Cambridge: Cambridge University Press, 1994); Ross Thomson, ed, *Learning and Technological Change* (New York: St. Martin's Press, 1993).

3. To succeed over the long run, both the computing and aerospace industries had to codify their innovation processes so that they became standard, routine ways of doing business. Good introductions to the literature on codification

include M. D. Cohen, R. Burkhart, G. Dosi, M. Egidi, L. Marengo, M. Warglien, and S. Winter, "Routines and Other Recurring Action Patterns of Organizations: Contemporary Research Issues," *Industrial and Corporate Change* 5 (1996), pp 653–698. See also R. Cowan and D. Foray, "The Economics of Codification and the Diffusion of Knowledge," *Industrial and Corporate Change* 6 (1997), pp 595–622.

4. James Cortada, *Before the Computer: IBM, NCR, Burroughs, Remington Rand and the Industry They Created, 1865–1956* (Princeton: Princeton University Press, 1993).

5. Roger E. Bilstein, *Flight in America, From the Wrights to the Astronauts,* rev ed (Baltimore: Johns Hopkins University Press, 1994); Jacob Vander Muelen, *Building the B–29* (Washington, D.C.: Smithsonian Institution Press, 1995).

6. See Stuart W. Leslie, *The Cold War and American Science: The Military–Industrial–Academic Complex at MIT and Stanford* (New York: Columbia University Press, 1993), pp 2–3.

7. See Walter A. McDougall, *The Heavens and the Earth: A Political History of the Space Age* (New York: Basic Books, 1985), for the early US and Soviet programs; Michael Rycroft, ed, *The Cambridge Encyclopedia of Space* (Cambridge: Cambridge University Press, 1990). On the German program, see Michael J. Neufeld, *The Rocket and the Reich: Peenemünde and the Coming of the Ballistic Missile Era* (New York: Free Press, 1995); Milton Lomask, *Robert Goddard: Space Pioneer* (Champaign, Ill: Garrard Publishing, 1972); J. D. Hunley, "A Question of Antecedents: Peenemünde, JPL, and the Launching of US Rocketry," in Roger D. Launius, ed, *Organizing for the Use of Space: Historical Perspectives on a Persistent Issue,* AAS Hist Series, vol 18, R. Cargill Hall, series ed (San Diego: Univelt, Inc., 1995), pp 1–31. On early pioneers, see Wernher von Braun and Frederick Ordway, III, *History of Rocketry and Space Travel* (New York: Crowell, 1966).

8. A standard introductory text is George Paul Sutton, *Rocket Propulsion Elements: An Introduction to the Engineering of Rockets,* 6th ed (New York: John Wiley & Sons, 1992).

9. This characterization comes from informal conversations in the early 1990s with engineers at Lockheed Martin, Denver, Colo.

10. The early symposia and conferences on reliability and quality assurance, sponsored by the military services, were almost completely monopolized by the missile designers. For example, see *Proceedings, National Symposium on Quality Control and Reliability in Electronics,* New York City, Nov 12–13, 1954, sponsored by the Professional Group on Quality Control, Institute of Radio Engineers and the Electronics Technical Cmte, American Soc for Quality Control (New York: Institute of Radio Engineers, 1955).

11. See, for example, *Bimonthly Summ Rpt No 36a, Suppl to Combined Bimonthly Summ No 36, the Corporal Guided Missile XSSM–A–17,* Aug 1, 1953, JPL, JPLA; and *Bimonthly Summ Rpt No 37a, Suppl to Combined Bimonthly Summ No 37, the Corporal Guided Missile XSSM–A–17,* Oct 1, 1953, JPL, JPLA.

12. Wilfrid J. Mayo-Wells, "The Origins of Space Telemetry," in Eugene M. Emme, *The History of Rocket Technology* (Detroit: Wayne State University Press, 1964), pp 253–270.

13. Author's experience, mid-1980s, Martin Marietta Corporation, spacecraft design.

14. For example, see note by the Secretariat, "Rpt on Launch of F6/1 Vehicle," ELDO/T(67)24, Paris, Sep 22, 1967, HAEUI ELDO Fond 2977.

15. See Stephen B. Johnson, "Insuring the Future: The Development and Diffusion of Systems Management in the American and European Space Progs," Ph.D. diss, University of Minnesota,

1997. For early failure rates, see pp 144–145 for JPL, p 106 for Atlas and Titan. See also Jacob Neufeld, *Ballistic Missiles in the USAF 1945–1960* (Washington, D.C.: Ofc of AF Hist, 1990), pp 217– 220. Current launcher reliability rates can be derived from data issued by the US Dept of Transportation; typically they hover between 85 and 95 percent.

16. Kent C. Redmond and Thomas M. Smith, *Project Whirlwind: A Case History in Contemporary Technology* (Bedford, Mass: MITRE Corporation, 1975); see also Atsushi Akera, "Calculating a Natural World: Scientists, Engineers, and Computers in the US, 1937–1968," Ph.D. diss, University of Pennsylvania, 1998, chapter 4, part I.

17. Author's experience, dependable system design.

18. Intvw, Robert Everett with author, Oct 1 and 13, 1998, USAF HSO.

19. During World War II and the Cold War, Johns Hopkins University was the home of the SRL and the Applied Physics Lab along with the Ops Research Ofc. For information on the SRL, see Paul N. Edwards, *The Closed World: Computers and the Politics of Discourse in Cold War America* (Cambridge: MIT Press, 1996), pp 218–219. For the postwar APL, see Michael Aaron Dennis, "'Our First Line of Defense': Two University Labs in the Postwar American State," *Isis* 85 (1994), pp 427–455.

20. Ellis Johnson, "Operations Research in the World Crisis in Science and Technology," in Charles D. Flagle, William H. Huggins, and Robert H. Roy, *Operations Research and Systems Engineering* (Baltimore: Johns Hopkins University Press, 1960), pp 28–57.

21. Vincent C. Jones, *Manhattan: The Army and the Atomic Bomb* (Washington, D.C.: Ctr for Military Hist, US Army, 1985); Lt Gen Leslie R. Groves, "The A-Bomb Program," in Fremont Kast and James Rosenzweig, eds, *Science, Technology, and Management* (New York: McGraw-Hill, 1963), pp 33–34, 39–40; Richard Rhodes, *The Making of the Atomic Bomb* (New York: Simon & Schuster, 1986), pp 486–496; Lillian Hoddeson, Paul W. Henriksen, Roger A. Meade, Catherine Westfall, *Critical Assembly: A Technical History of Los Alamos during the Oppenheimer Years, 1943–1945* (Cambridge: Cambridge University Press, 1993), pp 1–3 and chapter 8.

22. Interim rpt, von Kármán to H. H. Arnold, Gen of the AAF, "Where We Stand," Aug 22, 1945, in Michael H. Gorn, *Prophecy Fulfilled: "Toward New Horizons" and Its Legacy* (Washington, D.C.: AF Hist and Museums Prog, 1994); see intvw, Gen Bernard A. Schriever with author, Apr 13, 1999, USAF HSO. The V–2 project was not as well known to Americans despite von Kármán's rpt. It was obviously used as a model for Wernher von Braun's organization because many of his original team came with him to the US, ultimately forming the core of NASA's Marshall Space Flight Ctr in Huntsville, Ala.

23. There is evidence for this in the Burroughs Papers at CBI; see also Cortada, *Before the Computer.*

24. Verbal discussion, David Mindell, 1995; see also intvw, Dr. Ivan Getting with author, Oct 30 and Nov 6, 1998, USAF HSO.

25. David A. Mindell, "Automation's Finest Hour: Radar and System Integration in World War II," paper presented to Symposium on the Spread of the Systems Approach, Dibner Institute, Cambridge, Mass, May 3–5, 1996, pp 6–9; intvw, Getting, Oct 30, Nov 6, 1998; ltr, Ivan A. Getting to the author, Jun 18, 1999.

26. "Statement of Relationships between the Bureau of Ordnance, US Navy and the NDRC, OSRD, on the Development and Production of the Gunfire Control System Mark 56," reprinted in Ivan Getting, *All in a Lifetime: Science in the Defense of Democracy* (New York: Vantage Press, 1989), p 186.

27. Neufeld, *Ballistic Missiles in the USAF,* pp 226–228; see also John Lonnquest, "The Face of Atlas: Gen Bernard

Schriever and the Development of the Atlas Intercontinental Ballistic Missile 1953–1960," Ph.D. diss, Duke University, 1996, chapters 2–4. Schriever and Special Asst to the Secretary of the AF for R&D Trevor Gardner orchestrated the machinations that made ICBMs the top priority.

28. The early history of OR is described in Edwards, *Closed World;* Florence N. Trefethen, "A History of Operations Research," in Joseph McCloskey and Florence N. Trefethen, eds, *Operations Research for Management* (Baltimore: Johns Hopkins University Press, 1954); and in Air Ministry Publication 3368, *The Origins and Development of Operational Research in the Royal AF* (London: Her Majesty's Stationery Ofc, 1963). Recent articles on OR include Michael Fortun and Sylvan S. Schweber, "Scientists and the Legacy of World War II: The Case of Operations Research (OR)," *Social Studies of Science* 23 (1993), pp 595–642; Robin E. Rider, "Operations Research and Game Theory: Early Connections," in Roy E. Weintraub, ed, *Toward a History of Game Theory,* annual suppl to vol 24 of *History of Political Economy* (Durham: Duke University Press, 1992); Gene H. Fisher and Warren E. Waler, "Operations Research and the RAND Corporation," in Saul I. Gass and Carl M. Harris, eds, *Encyclopedia of Operations Research & Management Science* (Boston: Kluwer Academic, 1996); Stephen P. Waring, "Cold Calculus: The Cold War and Operations Research," *Radical History Review* 63 (Fall 1995), pp 28–51. Several OR papers were presented at the symposium, "The Spread of the Systems Approach," held at the Dibner Institute in Cambridge, Mass, May 3–5, 1996. They included: Erik Rau, "New Times, New Uses: Philip Morse, the Cold War, and the Proliferation of Operations Research"; David Jardini, "Out of the Blue Yonder: The Transfer of Systems Thinking from the Pentagon to the Great Society, 1961–1965"; Arne Kaijser and Joar

Tiberg, "The Establishment, Transformation and Diffusion of Operations Research in Sweden, 1945–1980"; and David Hounshell, "The Medium is the Message, or How Context Matters: The RAND Corporation Builds an Economics of Innovation, 1946–1965." Research on the history of OR in Britain also is ongoing at the University of Manchester.

29. Trefethen, "A History of Operations Research," pp 5–24; see also Rau, "New Times, New Uses"; Bruce L. R. Smith, *The RAND Corporation, Case Study of a Nonprofit Advisory Corporation* (Cambridge: Harvard University Press, 1966), chapters 1–2.

30. Hounshell, "The Medium is the Message"; Jardini, "Out of the Blue Yonder"; see also David R. Jardini, "Out of the Blue Yonder: The RAND Corporation's Diversification into Social Welfare Research, 1946–1968" Ph.D. diss, Carnegie Mellon University, 1996.

31. Fortun and Schweber, p 607.

32. For the practices of ops researchers, see the early OR texts and symposia, particularly Philip M. Morse and George F. Kimball, *Methods of Operations Research* (New York: John Wiley & Sons, 1951); C. West Churchman, Russell L. Ackoff, and E. Leonard Arnoff, *Introduction to Operations Research* (New York: John Wiley & Sons, 1957); Flagle, Huggins, and Roy, *Operations Research;* and Donald P. Eckman, ed, *Systems: Research and Design, Proceedings of the First Systems Symposium at Case Institute of Technology* (New York: John Wiley & Sons, 1961). Those researchers primarily applied existing mathematical techniques, simple theories from classical mechanics, and more complex ones from electromagnetic theory.

33. Eckman, *Systems,* p vi. Ramo was a cofounder of R-W Corporation, later the aerospace giant TRW.

34. Ellis A. Johnson, "The Executive, the Organization, and Operations Research," in McCloskey and Trefethen, *Operations Research for Management,* pp xvi–xvii.

35. *Ibid.,* p xi.

36. Arthur D. Hall, *A Methodology for Systems Engineering* (Princeton: D. Van Nostrand Company, 1962), p 12.

37. Trefethen, "History of Operations Research," pp 33–34.

38. Stuart Leslie, *The Cold War and American Science* (New York: Columbia University Press, 1993), pp 94–95.

39. Hall, *Methodology for Systems Engineering,* pp vii–viii, 20.

40. OR jnls included *Operations Research* and *Operational Research Quarterly.* The first text was Morse and Kimball's *Methods of Operations Research.* Systems engineers did not develop a journal for some time but they did create texts, starting with Harry M. Goode and Robert E. Machol, *Systems Engineering, An Introduction to the Design of Large-Scale Systems* (New York: McGraw-Hill, 1957) and Robert E. Machol, Wilson P. Tanner, Jr., and Samuel N. Alexander, *System Engineering Handbook* (New York: McGraw-Hill, 1965), which became the "bible" of sorts for systems engineers in the late 1960s.

41. See, for example, Dennis Lock, *Project Management,* 6th ed (New York: John Wiley & Sons, 1996). The Phase A–D terminology is used by NASA; phase 1–4 terminology is used by the DOD.

42. Leslie, *The Cold War and American Science;* Roger L. Geiger, *Research and Relevant Knowledge: American Universities Since World War II* (New York: Oxford University Press, 1993); Daniel J. Kevles, *The Physicists: The History of a Scientific Community in Modern America* (Cambridge: Harvard University Press, 1971).

43. For a thorough and critical discussion of concurrent versus sequential procurement in the military, see Michael E. Brown, *Flying Blind: The Politics of the US Strategic Bomber Program* (Ithaca: Cornell University Press, 1992). For an insightful look at "concurrency" as used by its primary mouthpiece, Gen Bernard Schriever, see Lonnquest, "Face of Atlas."

44. US Cong, House Cmte on Armed Services, *Hearings, Investigations of National Defense Missiles,* 85th Cong, 1958; US Cong, House Cmte on Armed Services, *Hearings, Weapons System Management and Team System Concept in Government Contracting,* 86th Cong, 1st Sess, 1959; US Cong, House Cmte on Government Ops, *Hearings, Organization and Management of Missile Progs,* 86th Cong, 1959; US Cong, House Cmte on Government Ops, Subcmte on Military Ops, *Organization and Management of Missile Progs,* 86th Cong, 1st Sess, rpt no 1121.

45. A standard text describing the functioning of government bureaucracy is Harold Seidman and Robert Gilmour, *Politics, Position, and Power: From the Positive to the Regulatory State,* 4th ed (New York: Oxford University Press, 1986).

Chapter 2

Building the Air Force of the Future

> If the physicists, the chemists, the mathematicians, and the engineers could combine to build an atomic bomb, why could not the same kinds of groups, working in concert, solve other problems, both military and civil? The concept of the multidisciplinary approach was utilized in World War II, and it was only natural that the techniques thus devised should carry over.
>
> P. Stewart Macaulay, 1960[1]

Relying on complex devices to wage war or even to be present in the air, the U.S. Air Force always has been highly dependent on technology. World War II and the Cold War made this dependence even more complete with the development of radar to detect aircraft from a distance far beyond human sight, jet engines to propel humans through the air more rapidly than before, nuclear weapons to unleash unprecedented destruction, and ballistic missiles to transport nuclear weapons to any location on the globe within minutes. These new wonder weapons demonstrated the utility of scientific and technological research for military power. Military leaders could not afford to ignore the possibility of technological revolution emanating from some new scientific or technological breakthrough. For many Air Force officers, the only question was how to maintain and improve the close relationships built during World War II.

New technology provided opportunities for military officers with a technical bent. Allied with scientists and research engineers, these officers promoted the "Air Force of the future," in contrast with the traditional "Air Force of the present." Only through the promotion of new technology using wide-ranging research and fast-paced development would the Air Force maintain its critical technological edge over its Communist enemy.

By separating research and development from the support of current operations, these officers brought technological development to the forefront of Air

Force concerns. Then, by creating new methods to integrate technologies into novel weapons systems, they harnessed the potential of new technology into powerful tools for the projection of power. In so doing, they also brought into being new processes, organizations, and niches for technologically minded officers, scientists, and engineers.

Army Aircraft Procurement Through World War II

Military aircraft procurement involved close interactions between government and industry from its inception in the first decade of the twentieth century. Because the Army did not create its own arsenal to develop aircraft, contractual relationships between the Army Air Corps and aircraft industry governed military aircraft development. The Army Signal Corps ordered its first aircraft from the Wright brothers in 1908 using an incentive contract that awarded higher fees for a higher-speed aircraft.[2] Army evaluation and testing of aircraft began near the Wright's plant in Dayton, Ohio, and grew into the Air Corps' primary facility for testing aircraft, equipping them with weapons, and contracting with the aircraft industry.

The Army's seemingly lackadaisical attitude toward the use of aircraft troubled some officers when World War I began and U.S. involvement became increasingly likely. While the European powers rapidly developed aircraft for significant military purposes, the U.S. Army held aircraft development and pilot training as a low priority. With few aircraft and little capability, the United States compared poorly with the Europeans, who deployed hundreds of high-performance aircraft. To alleviate this problem, in 1915 Congress created the National Advisory Committee for Aeronautics (NACA) to promote aircraft research, evaluation, and development for the Army, the Navy, and the aircraft industry. Engineers at NACA's facility at Langley Field in Hampton, Virginia, concentrated on testing and evaluating aerodynamic structures and aircraft performance, using new wind-tunnel facilities to test fuselages, engine cowlings, propeller designs, and pilot-aircraft controllability.[3]

Government procurement regulations hampered the efforts both of aircraft companies and Army Air Corps procurement officers. During the First World War, large profits made by some military contractors using negotiated cost-plus-percentage-of-cost contracts made Congress wary of all negotiated contracts. Members of Congress wanted to ensure competition through competitive bids for fixed-price contracts, and by separating design and development from production contracts. The Army Air Corps could let small contracts for experimental aircraft and designs, which by law the government acquired. Procurement officers then released the design for production bids. This led to cases where one company would design an aircraft at substantial expense to itself, but another company that did not have to recoup the large up-front investment could acquire the design free of charge from the government, submit a lower bid, and manufacture the aircraft, thus leaving the innovative company with a huge loss. Because the aircraft industry made its money on manufacturing aircraft in (hopefully) large production runs, aircraft companies responded by avoiding Army Air Corps R&D.[4]

Congress passed new legislation in 1926 to remedy these problems. The Air Corps Act of that year required design competitions for production, in which the Air Corps specified performance characteristics for the aircraft instead of a particular design. An evaluation board then would evaluate the resulting designs—just the paper design, not a prototype—using a public-record numerical rating scheme. By the late 1930s the Air Corps used three boards, one each for pilots, engineering officers, and tactical officers. The winner won a fixed-price production contract.[5]

Air Corps procurement officers found the new procedures with their design competitions to be unworkable. First, companies competed with paper designs bearing little resemblance to the final product. It took a great deal of time and money to develop an aircraft to the point where it could be tested, at which point the Air Corps sometimes found it inadequate. Aircraft manufacturers also disliked the new regulations because they could not accurately estimate production costs based on a paper design. They lost money on the fixed-price contract even if they won the design competition.[6]

Constrained by one set of regulations, Army procurement officers resorted to other regulations to award the production contract to the designing firms with a solid track record. Congress saw this as evidence of a conspiracy and took steps to restrict the practice. Astute maneuvering by Assistant Secretary of War H. H. Woodring avoided that result. In 1934 he promised Congress that production contracts would be awarded competitively but with a new twist. Firms would have to submit an actual aircraft, not simply drawings. This restricted production contracts to large, financially viable companies that could afford to submit an aircraft for its production bid. The government also imposed standardization across designs and required aircraft companies to purchase government-furnished equipment for instruments, communications, and armaments.[7]

As the war in Europe loomed in 1940, the Roosevelt administration took a number of steps to mobilize the research and development capabilities of the country. Vannevar Bush created the National Defense Research Committee (NDRC) and later the Office of Scientific Research and Development (OSRD).[8] Using NACA as a model, Bush organized NDRC as an independent research organization with military and scientific representatives. NDRC then let research contracts for critical defense problems. The OSRD expanded these functions into technology development.[9]

Bush favored voluntary associations between scientists and engineers on the one hand and the government and military on the other. This approach discouraged direct government or military control that he feared might squelch innovative ideas and, not coincidentally, the political influence of R&D experts. His "associationist" strategy placed scientists and engineers on an equal level with their bureaucratic superiors. His close relationship with President Franklin Roosevelt and the wartime emergency aided the triumph of this viewpoint over military leaders, who wanted firm control over the scientists, and "reform liberals," who wanted stronger government control over both military and civilian R&D.[10]

Just as important, Congress legalized negotiated aircraft production contracts on a cost-plus-fixed-fee basis in 1940. With a flood of funding and a goal of building fifty thousand aircraft, the Army Air Corps immediately began negotiating contracts with industry. Under the prior competitive bidding process, procurement officers did not need to understand the financial details of the aircraft manufacturers' bid because underbids cost the manufacturer money, not the government. Under cost-plus-fixed-fee arrangements, however, cost overruns would be the government's problem. The Air Corps Procurement Branch grew rapidly to collect information and negotiate with contractors to assess the validity of cost charges and to determine a fair profit.[11]

The Army Air Forces* not only changed procurement practices but also diffused "best techniques" through the aircraft industry. One example was the spread of subcontracting from the prime contractor to lower-tier subcontractors, pioneered on a large scale by Lockheed. Another was the diffusion of "series" and "block" design and production methods, a simple numbering scheme to label major and minor changes. For example, a B-24J-15 was a minor modification of the B-24J-14 whereas the B-24K-1 was a major change from the B-24J. The series–block nomenclature allowed production of standardized items with well-tracked modifications and variations. Procurement officers ensured industrial use of production flow-charts on a consistent basis, as well as weekly reports from the contractor to the Army Air Forces. A third significant organizational innovation spread by the Air Corps was a committee system for contractor coordination initially pioneered by Boeing, Douglas, and Vought, contractors for the B–17.[12]

Wartime experience with negotiated contracts formed the basis for the contractual relationships that later typified the Cold War. Government officials became both partners and controllers of the aircraft industry in a way unimagined before World War II with expanded procurement organizations that made the federal government a formidable negotiator. These close working relationships between the Army Air Forces and industrial contractors, and the government's new ability to monitor industry, formed the basis for postwar developments.

Forming Organizations to Communicate with the Technologists

During the Second World War, scientists vastly increased the fighting capability of both Allied and Axis powers. The atomic bomb, radar, jet fighters,

---

*The Air Corps was renamed the Army Air Forces in World War II.

Maj. Gen. Curtis E. LeMay, Brig. Gen. Emmett O'Donnell, Gen. Henry H. Arnold, and Lt. Gen. Barney M. Giles.

ballistic missiles, and operations research methods applied to fighter and bomber tactics all had significant impact on the war. Recognizing the contributions of scientists during the war, Gen. Henry H. ("Hap") Arnold, the commander of the Army Air Forces, was one of a number of leaders who advocated maintaining the partnership between military officers and scientists after the war ended. Even as he planned for postwar demobilization, Arnold took measures to continue this collaboration. His plans led to creation of several organizations that solidified the partnership between technically minded Air Force officers and the community of scientific and technological researchers.

In 1944 Arnold met briefly with eminent aerodynamicist Theodore von Kármán and asked him to assemble a group of scientists to evaluate German capabilities and to study the Army Air Forces' postwar future. Among the group's recommendations were the establishment of a high-level staff position for R&D, establishment of a permanent board of scientists to advise the Air Forces, and creation of better means to educate AAF officers in science and technology.[13] The AAF acted first to maintain the services of von Kármán and his scientific colleagues. Supported by Generals Arnold and LeMay, the Air Force established the Scientific Advisory Board in June 1946 as a semipermanent adviser to the AAF staff. The deputy chief of Air Staff for research and development acted as liaison between the board and the Air Staff.[14]

Arnold recognized that establishing an external board of scientists would do little to change the Army Air Forces unless he also created internal positions that would act as bridges and advocates for scientific ideas. He established the position of scientific liaison on the Air Staff under the assistant chief of staff, materiel, and elevated his protégé Col. Bernard Schriever into the position in 1946. Schriever had known Arnold since 1933, when as a reserve officer he was a bomber pilot and maintenance officer serving under Arnold. Schriever's mother became a close friend of Arnold's wife and that led to a lifelong friendship between the two families. When Schriever married the daughter of another AAF general, Arnold gave away the bride in a ceremony performed in Arnold's living room. Arnold encouraged Schriever to take a full commission and he did so prior to World War II. Schriever served with distinction in the Pacific theater and his work in logistics brought him into contact with Air Materiel Command (AMC) officers at Wright Field. After the war, Arnold moved him to the Pentagon. As the scientific liaison, Schriever had a hand in the AAF's efforts to create an infrastructure for R&D. Those efforts included test facilities at Cape Canaveral, Florida, and in the Mojave Desert north of Los Angeles, and research centers in Tennessee and at Hanscom Field near Boston. In so doing, he worked closely with the SAB, an association that would have far-reaching consequences.[15]

Arnold also took measures to create the first so-called think tank, the RAND Corporation. Initially the brainchild of engineer Frank Collbohm of Douglas Aircraft Company and Arnold's special consultant Edward Bowles, Project RAND was charged with investigating future weapons for the Army Air Forces. Arnold had research funding that he allocated to the project, and in March 1946 Project RAND came into official existence as a contract with the Douglas Aircraft Company to perform research on intercontinental warfare. Because Collbohm and his Douglas Aircraft Company colleagues had worked on improving B–29 operations during the war using operations research methods, Project RAND started out by applying operations research to the much larger problem of intercontinental warfare.[16] In the next year, Project RAND became the nonprofit RAND Corporation, as Douglas Aircraft representatives feared that the project would create an "insider" position that could jeopardize hardware contracts with the Air Forces.

During World War II, operations researchers focused on the tactical operations of existing weapons. In contrast, RAND's research investigated the potential value of future systems. RAND researchers used many of the techniques developed by operations researchers, extended by their best estimates of future technological and operational trends. They called this future-oriented operations research "systems analysis."[17] Like operations research, systems analysis required mixed teams of mathematicians, scientists, engineers, managers, economists, and military personnel.[18]

Frank R. Collbohm, president of RAND Corporation, 1958–1961. He suggested that RAND personnel move to Washington, D.C., to assist with systems analysis for the Air Force DCS/D. These personnel later formed the core of the ANSER Corporation. Collbohm also promoted RAND's Systems Research Laboratory experiments as potential solutions for air defense training.

The Army Air Forces' ongoing battle to become an independent service also led to further associations between scientists and the AAF. Among the many provisions of the National Security Act of 1947 that established the Department of Defense and the United States Air Force was the creation of the Research and Development Board (RDB), a committee consisting of two representatives from each service with a civilian chairman. The RDB extended the Joint Research and Development Board, established in 1946 to coordinate research efforts between the Army and the Navy. It operated like its predecessor but now included the Air Force. The RDB coordinated scientific research related to national security among the services, advised the services and the secretary of defense about scientific trends, and formulated policy for the interactions of military R&D with outside organizations. To do its work the RDB established specialized panels to review and report activities in their specific disciplines.[19]

Because it did not have directive authority, the RDB was only marginally effective at coordinating the R&D activities of the services. For new technologies such as missiles, the board could do little more than provide a forum in which arguments about missions and technological approaches could take place. It could not prevent the services from duplicating missions, functions, and programs. However, the board was a reasonably effective communications device whereby military officers, scientists, and research engineers could learn of each others' concerns, programs, and approaches.[20]

Knowing about scientific capabilities and interests was one thing but harnessing them for military purposes was another. During World War II, the Office of Scientific Research and Development had the authority to negotiate contracts with industry, academia, and individuals. Unless Congress extended that authority after the war, the military's capability to control scientists and their new technologies would dramatically decrease. Fortunately for the military, the Procurement Act of 1947 extended the military's wartime authority and tools into peacetime, including the formerly controversial negotiated contract mechanism.

The importance of the Procurement Act should not be underestimated for it perpetuated government use of the cost-plus contracts, an act that had several important ramifications. First, the cost-plus contract reduced individual and institutional risk. Where high risk was inherent, as in research and development, this drew profit-making corporations and universities into government-run activities. Second, to reduce government risk the cost-plus contract required maintaining a government bureaucracy sufficient to monitor contractors. Third, the cost-plus contract turned attention away from cost concerns to technical issues. This "performance-first" attitude led to higher costs but also to a faster pace of technical innovation that occasionally led to radical technological change. Such changes frequently led to the creation of economically significant enterprises in both public and private sectors. Last, the cost-plus contract provided some military officers with the means to promote technological innovation along with their own careers.[21]

Taking control of some but not all Army organizations, Air Force officers found themselves without the "arsenal" capability typical of the Army. Without the technical expertise of the Navy's design bureaus or the Army's arsenals, the Air Force had to build its own capability and to depend on civilian capabilities in industry and academia. Turning their disadvantage into an opportunity, Air Force officers expanded their contractual relationships with industry and with scientists. This expansion had two important political consequences. With scientific prestige at high levels after the development of radar and atomic weapons during World War II, scientists placed in positions of power throughout the defense establishment became powerful advocates for the Air Force. Similarly, the Air Force's dependence on industry meant that "pork politics" worked to the Air Force's advantage over the Army or Navy. Air Force contracts that were spread judiciously through important congressional districts could and did lead to political pressure favoring Air Force programs.

The Navy had a significant initial advantage in its relationships with scientists through its Office of Naval Research (ONR). Recognizing a postwar opportunity with the dissolution of the Office of Scientific Research and Development, the ONR provided a significant share of all basic research funds prior to the intrusion of the Air Force and to creation of the National Science Foundation in 1950.[22]

The Air Force's entry into research was more difficult. It had to compete not only with the ONR but also with the activities of the Atomic Energy Commission, the RDB, and NACA.[23] In addition, its initial research activities and laboratories were fragmented and uncoordinated.

Dr. Theodore von Kármán, chairman of the Scientific Advisory Board; Brig. Gen. Donald L. Putt, director of R&D for the deputy chief of staff/materiel; and Dr. Albert E. Lombard, Jr., head of the research division under General Putt.

Seeing the success of the ONR and not wanting the Navy to control all R&D, Theodore von Kármán proposed to the Air Staff in 1947 that the Air Force establish its own basic research organization. Lt. Gen. Benjamin Chidlaw, the commander of Air Materiel Command, successfully argued that this organization should be housed in his facilities at Wright Field where the bulk of Air Force R&D resided at the time. Consequently, AMC established the Applied Research Section of the Engineering Division in February 1948. In February 1949 the section was renamed the Office of Air Research and was moved out of and parallel to the Engineering Division. The office tried to hire a few scientists but with limited budget and substantial administration they were largely unsuccessful.[24]

The Air Force inherited a number of research, development, and testing facilities from the Army. These included facilities at air bases such as Kirtland, Holloman, Muroc, Clinton, and Griffiss, and two laboratories, the Watson Laboratories in New Jersey and the Cambridge Research Laboratories near Boston. Because funding for R&D facilities often was intermixed with funding for operations of the bases, management of R&D was frequently compounded with other issues. For its first few years, the Air Force spent a great deal of time simply devising organizational means to understand and coordinate the efforts of its diverse facilities.[25]

Typical of the challenges was the problem of requirements. In the summer of 1949, Vice Chief of Staff Gen. Muir S. Fairchild requested that Dr. Edmund P. Learned of the Harvard Business School review the Air Staff organization. Learned commented on a number of issues in his report of July 29. His biggest concern was with the fragmentation of Air Force efforts related to requirements. He found that requirements originated from many places and that no formal mechanism existed to determine priorities or select among them. In addition, coordination between Headquarters staff agencies and AMC was lacking without a central organization. He recommended creating a new position to coordinate and prioritize requirements to be used to create new technologies and systems. Fairchild wasted no time in responding. On the fifth of August he created a new Directorate of Requirements under the deputy chief of staff, operations.[26]

Despite the creation of a research office in AMC, an increasing number of military officers believed that Air Materiel Command was not furthering research and development with sufficient vigor. The controversy revolved around technologically oriented officers who promoted the "Air Force of the future" versus the traditional pilots who focused on the "Air Force of the present." The advocates of the future Air Force had powerful allies in General Arnold and in Lt. Gen. Donald Putt, a longtime aircraft procurement officer from Wright Field. Putt had been a student of von Kármán at the California Institute of Technology, and in the late 1940s was director of research and development under the chief of staff, materiel in the Air Force Headquarters staff.[27] Figure 2 shows Putt's dual position on the Air Force Headquarters staff.

Putt and an energetic group of colonels under him discussed how to improve Air Force R&D, which in their opinion languished in AMC. As budgets shrank after the war, AMC gave higher priority to maintaining operational forces than to R&D, thus leading to large R&D budget cuts. This trend concerned many Air Force officers as well as members of the Scientific Advisory Board. Putt and his enthusiastic young officers plotted how the SAB could aid their cause.[28]

Capitalizing on an upcoming meeting of the SAB in the spring of 1949, Putt asked Chief of the Air Staff Gen. Hoyt S. Vandenberg to speak to the board about Putt's concerns. Vandenberg agreed but only if Putt would write his speech. This was the opportunity that Putt and his young protégés were looking for. Putt asked Col. Ted Walkowicz, the military secretary to the SAB, to write the speech. Walkowicz included "a request of the Board to study the Air Force organization to see what could be done to increase the effectiveness of Air Force Research and Development." Putt "rather doubted that Vandenberg would make that request." Fortunately for Putt, at the last minute Vandenberg backed out and had his deputy, General Fairchild, appear before the board. Fairchild, a supporter of R&D, read the speech all the way through, including the request. Putt had already warned SAB chairman von Kármán what was coming so von Kármán quickly accepted the request.

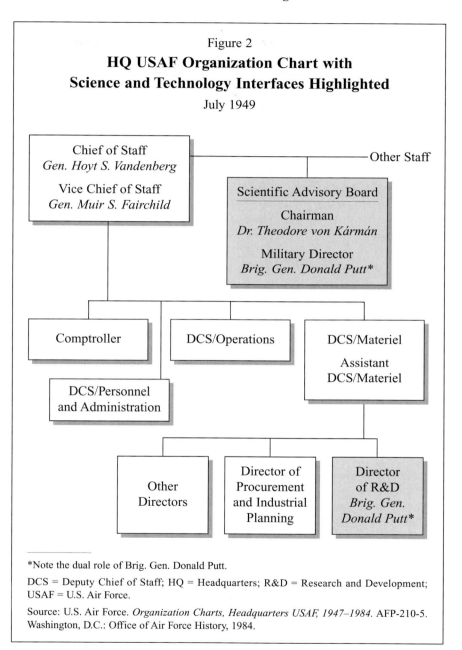

Figure 2
**HQ USAF Organization Chart with
Science and Technology Interfaces Highlighted**
July 1949

*Note the dual role of Brig. Gen. Donald Putt.

DCS = Deputy Chief of Staff; HQ = Headquarters; R&D = Research and Development;
USAF = U.S. Air Force.

Source: U.S. Air Force. *Organization Charts, Headquarters USAF, 1947–1984.* AFP-210-5.
Washington, D.C.: Office of Air Force History, 1984.

Louis Ridenour became chief scientific adviser for the DCS/D in 1950.

Putt and his colleagues knew that this was only the first step in the upcoming fight; they also had to ensure that the report actually would be read. Putt's group carefully handpicked the SAB committee to include members who had credibility in the Air Force. One was Louis Ridenour, well known for his work on radar at MIT's Radiation Laboratory. More important was the inclusion of James Doolittle, famed AAF bomber pilot and pioneer aviator who was also Vandenberg's close personal friend. Putt persuaded Doolittle to go on a duck-hunting trip with Vandenberg after Ridenour and von Kármán presented the study results to Vandenberg and the Air Staff. Putt later commented that "this worked perfectly," gaining the chief's ear and favor. Putt's group also coordinated a separate Air Force review to assess the results of the scientific committee. After handpicking its members as well, and ensuring coordination with Ridenour's group, Putt noted that "strangely enough, they both came out with the same recommendations."[29]

The Ridenour Report charted the Air Force's course over the next few years. It recommended creating a new command for R&D, a new graduate study program in the Air Force to educate officers in technical matters, and improvement of career paths for technical officers. The report also recommended creating a new general staff position for R&D separate from logistics and production and a centralized accounting system to track R&D expenditures more satisfactorily. On January 3, 1950, after a few months of internal debate, General Fairchild approved the creation of Air Research and Development Command (ARDC), which separated the R&D functions from Air Materiel Command. Along with ARDC, Fairchild approved creation of a new Air Staff position, the deputy chief of staff, development (DCS/D).[30]

With the establishment of ARDC and the DCS/D, the Air Force completed development of its first organizations to cement the ties between technically minded military officers and scientific and technological researchers. In theory these new organizations, which also included RAND, the RDB, and the SAB, would make the fruits of scientific and technological research available to the Air Force. The three organizations coordinated Air Force efforts with the help of the scientists and engineers, similar to Bush's wartime OSRD and its "associationist" model. ARDC and the DCS/D would attempt to centralize and control the Air Force's R&D efforts. To accomplish this, they would not only have to create a new organization but also would have to work out effective relationships with its occasionally jealous parent organization, Air Materiel Command, and its Air Staff counterpart, the deputy chief of staff, materiel (DCS/M).

Development Planning and the Organization of the DCS/D

Although the DCS/D and ARDC came into official existence rather quickly, with approval on January 3 and activation on January 23, 1950, General Fairchild expected the transition to be difficult because both organizations had to assume responsibilities formerly held by the DCS/M and AMC. To further confuse the picture, the DCS/D had to learn to confine itself to "executive" functions such as planning and policy formation, leaving day-to-day activities to ARDC. To handle the anticipated problems, Fairchild appointed General Doolittle (retired) as a special assistant to guide the transition period and arbitrate any conflicts. Doolittle, who had helped convince Chief of Staff Vandenberg of the necessity for an R&D organization, was well respected by all parties and interested in success. Fairchild expected that he would be able to arbitrate any disputes. Louis Ridenour became chief scientific adviser for the DCS/D in the summer of 1950, and Vandenberg assigned the critical post of DCS/D to the outspoken former commander of Air Defense Command, Gen. Gordon Saville.[31]

Saville, at the time assigned as the director of requirements under the deputy chief of staff, operations (DCS/O), was a prickly maverick dedicated to developing air defense, in contrast to the "bomber" mentality of many of his fellow officers. He proposed moving the director of requirements from under the DCS/O to the new DCS/D. Although his idea was opposed by many on the Air Staff, Saville carried the day and placed this office of the "flying Air Force" into the world of R&D. General Nelson held the director of requirements position. Also serving under Saville was the assistant for evaluation, a role temporarily filled by Colonel Schriever; the assistant for development programming under Brig. Gen. Donald Yates; and the director of R&D, General Putt.[32]

Through his career working on air defense, Saville met many of the scientists and engineers involved with the development of radar and other electronic equipment. One of them was Dr. Ivan Getting, the former Radiation Laboratory physicist and in 1949 the chairman of the Electronics Committee of the Research and Development Board. Saville asked Getting to join him in the DCS/D office in January 1950, but only in August did Getting accept the position after the outbreak of the Korean War.[33]

Getting filled the position of assistant for evaluation, tracking scientific and technological developments and connecting those new developments to future

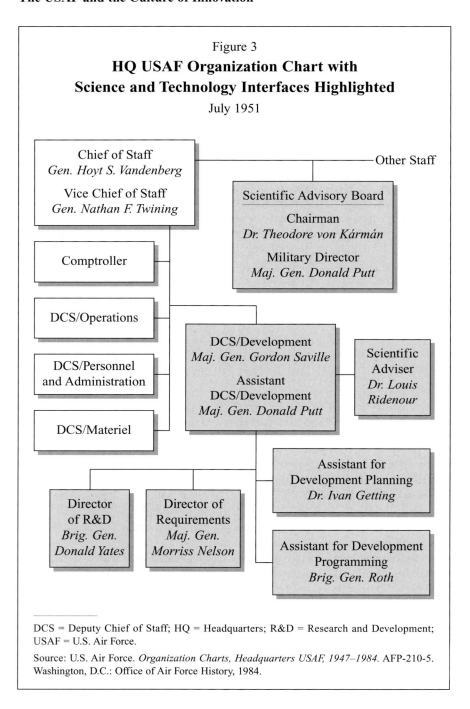

Figure 3

**HQ USAF Organization Chart with
Science and Technology Interfaces Highlighted**

July 1951

DCS = Deputy Chief of Staff; HQ = Headquarters; R&D = Research and Development; USAF = U.S. Air Force.

Source: U.S. Air Force. *Organization Charts, Headquarters USAF, 1947–1984.* AFP-210-5. Washington, D.C.: Office of Air Force History, 1984.

Air Force strategy and operations. Assisting him was his deputy Colonel Schriever, the former scientific liaison. Getting assigned Schriever the responsibility for strategic matters, and other assistants handled tactical issues, air defense, guidance, and so on.[34]

One major task of the assistant for evaluation was to make plans for developing future Air Force systems and technologies. This effort began in August 1950 when General Yates, the assistant for development programming, requested that Getting's office begin "immediate preparation of development planning objectives which would point out the goal of research and development effort in relation to the strategic and operational plans of the Air Force. Such objectives would take into account future combat conditions, possibly enemy and U.S. capabilities, major requirements, etc." These objectives would then form the basis for allocation and programming of R&D funding through Yates' office.[35]

To generate such development objectives, Getting's office could look in two directions: "operations pull" from the needs of current operations through the Directorate of Requirements, or "technology push" from the expertise of scientists and technologists. For the latter, they had ready access to the Air Force's specialists in analysis of future systems, the RAND Corporation. Unfortunately, RAND had just embarrassed itself with its recently completed analysis of strategic offensive warfare.

Since its inception in 1946, RAND researchers had been progressively developing systems analysis methods in an attempt to create a "science of warfare." They had concentrated on specific component studies that would be necessary for large-scale analyses of strategic offense and defense. When the United States detected radiation from a Soviet atomic bomb test in September 1949, Air Force officers enthusiastically supported expansion of RAND's efforts, including its large-scale offense and defense studies. RAND scientists pressed rapidly forward on its offensive study and presented their results on March 1, 1950. Unfortunately for them, their first major product met with massive criticism from the Air Force.

RAND recommended that the most cost-effective way to rain destruction from the air on the Soviet Union was to flood Soviet skies with fuel-efficient turboprop bombers instead of fewer high-performance bombers. This recommendation offended almost everyone in the Air Force for many very good reasons. First, the study quantified the value of human life just as it did machinery and consequently had no moral or other difficulties with a strategy that would sacrifice many pilots for better destructive efficiency. This did not sit well with an Air Force composed primarily of current and former pilots. Second, the study assumed static force levels and hence tried to maximize cost effectiveness over high performance. Air Force officers assumed that with appropriate justification they could secure more funds to get both quantity *and* quality. Third, RAND's bomber strategy contradicted actual Air Force plans for how bombers would be based and operated. Fourth, the study assumed large quantities of fissile materials

Gen. Gordon Saville was a prickly maverick who was dedicated to developing air defense.

for nuclear bombs when these quantities simply did not exist. RAND's quantification error was understandable because the researchers did not have access to the highly classified nuclear program. Overall the study seemed bent on alienating virtually every group in the Air Force and caused RAND acute embarrassment. Although RAND developed mathematical methods that advanced the state of the art, the assumptions that went into the study undermined their efforts. From that time onward, RAND researchers scaled back their efforts and better grounded their assumptions.[36]

In November 1949, just prior to his selection as DCS/D, General Saville had visited RAND to encourage the organization to accelerate its work on air defense. By April 1950 RAND researchers had developed a grand plan for this large study, but the criticisms of their earlier study caused them to scale back their efforts drastically and focus less on methodology and more on concrete and realistic recommendations. The resulting study was much smaller, completed in the spring of 1951, and published as a RAND report in October of that year. This study avoided the gross errors of RAND's offense study and formed the basis for air defense work in the DCS/D office.[37]

Although this success showed that RAND's scientists could perform tasks useful to Air Force planning, RAND got into further trouble on other grounds. In general, RAND's leaders wanted to stay out of situations where they would judge contractors, but the Air Force asked them for help in assessing the contract bids for the F–102 program. Insisting on a time-consuming academic analysis, RAND researchers eventually chose Convair as the contractor. Unfortunately, the Air Force could not wait for RAND's slower time line and had already selected North American. Concerned about a possible scandal and about RAND's apparent lack of responsiveness to the Air Force's needs, General Vandenberg asked the head of evaluation, Ivan Getting, to talk with RAND's leaders and board of directors. Vandenberg believed Getting would be more effective in communicating with RAND because, as he put it, "they'll just say I do not understand the scientific approach, but you are a longhair and can explain it to them."[38]

Getting met with RAND's director, Frank Collbohm, in Santa Monica and then met with Rowan Gaither, the chairman of RAND's board of trustees. On the basis of discussions had at those meetings, in May 1951 Collbohm suggested that RAND and the DCS/D office place personnel in each other's offices on a rotational basis so as to improve working relationships and mutual understanding. In July 1951 four RAND personnel relocated to the DCS/D office under the assistant for development planning (formerly the assistant for evaluation). Eventually the Washington representatives of RAND formed the core of the ANSER Corporation, which spun off from RAND to assist the Air Staff.[39]

Based on Yates' request for development planning objectives in the fall of 1950, the office of the DCS/D described a new set of procedures to establish Air Force technology goals and to translate those goals first into requirements and then into funded development programs. The assistant for development planning would create "development planning objectives" or goals for new technology development. The director of requirements then would translate the objectives into specific requirements known as "general operational requirements" or GORs. The director of R&D would translate the requirements into specific orders known as "development directives." Finally, the assistant for development programming would allocate funds based on those prioritized directives.[40] Figure 4 depicts this stepwise process and the interoffice coordination that each step required.

H. Rowan Gaither, Jr., chairman of RAND Corporation and first chairman of the board of trustees for MITRE Corporation, 1958–1961. Gaither helped organize and finance both corporations.

The name change from the "assistant for evaluation" to the "assistant for development planning" in the spring of 1951 signified the evolution and increasing importance of this office and its new procedures. Charged with developing the long-term objectives of the Air Force's technology programs, Getting and his deputy Schriever looked both to RAND's systems analysis to assess long-term "technology push" and to the Air Force's ongoing operations to assess current and near-term needs. When General Saville retired in the spring of 1951, Getting

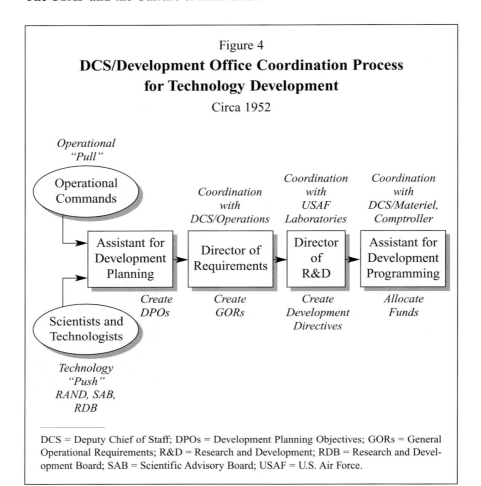

Figure 4

DCS/Development Office Coordination Process for Technology Development

Circa 1952

Operational "Pull"

Operational Commands

Coordination with DCS/Operations

Coordination with USAF Laboratories

Coordination with DCS/Materiel, Comptroller

Assistant for Development Planning → Director of Requirements → Director of R&D → Assistant for Development Programming

Create DPOs *Create GORs* *Create Development Directives* *Allocate Funds*

Scientists and Technologists

Technology "Push"
RAND, SAB, RDB

DCS = Deputy Chief of Staff; DPOs = Development Planning Objectives; GORs = General Operational Requirements; R&D = Research and Development; RDB = Research and Development Board; SAB = Scientific Advisory Board; USAF = U.S. Air Force.

left to take an executive post at Raytheon, Schriever took his place as assistant for development planning, and Putt became the assistant DCS/D.[41]

Contrary to the earlier practice of asking the operational commands what their needs were, Schriever required an analysis of future technologies, strategies, and objectives using systems analysis along with requirements from current operations to establish development planning objectives for future systems. These new, controversial methods led to battles between Schriever, who represented the scientists and the long-term future of the Air Force, and officers with a shorter-term view of their needs, like the powerful vice chief of staff, Gen. Curtis E. LeMay, who vigorously fought some of Schriever's recommendations. As Schriever put it, he didn't win too many of these battles, except on missiles, where there were fewer entrenched interests.[42]

The office of the DCS/D expanded its planning capabilities through the assistant for development planning but it had to let go of its day-to-day operations responsibilities, and the staff did so rather reluctantly. Personnel there worked on a description of their tasks and responsibilities and in August 1950 submitted it to the rest of the Air Staff for comments and revisions. The revisions came back with a consistent theme: change your focus from "hardware" to high-level policy and guidance. This feedback prompted a flurry of activity and a high-level staff meeting among the DCS/D, DCS/M, ARDC, and AMC, chaired by arbitrator James Doolittle. One major result of that meeting was the decision to transfer the day-to-day R&D management responsibilities by moving knowledgeable DCS/D personnel to ARDC.[43]

The transfer of personnel gave ARDC more capability, but it did not fully resolve the confused responsibilities between ARDC and the DCS/D. There were no new initiatives to delineate responsibilities until after Saville's retirement in July 1951. At that time General Yates became Acting DCS/D, and the following month he brought in Brig. Gen. James McCormack, Jr., as a special assistant to investigate and report on the organization of Air Force R&D. McCormack had been a Rhodes Scholar at Oxford and held a master's degree in civil engineering from MIT. His impressive background and obvious intelligence brought credibility to his work. Coming amid continuing complaints by ARDC, McCormack's report was to become an influential blueprint for action.[44]

Despite the transfer of personnel, ARDC complained that the DCS/D office was "reluctant to concede the Air Research and Development Command proper position in the Research and Development management cycle" primarily because the Research and Development Command was not properly staffed with a sufficient number of qualified officers. McCormack saw things a bit differently, believing that the organizations needed clear delineation of responsibilities. The problem was that the two organizations had distinctly different views of what their responsibilities should be. McCormack suggested that the DCS/D and ARDC separately define the responsibilities of each organization and then try to work out the differences. Interestingly, McCormack's analysis of the "line of demarcation" between the two organizations may have been the first use of the term "interface" in an organizational rather than technical context.[45]

McCormack's suggestion led to a high-level meeting once again chaired by Doolittle on December 18, 1951, among the newly appointed Acting DCS/D Putt, Director of R&D Yates, ARDC Commander Lt. Gen. Earle E. Partridge, and a number of staff officers. In part, the problem was that prior to the creation of ARDC, the research and development laboratories of the Engineering Division, Air Materiel Command, reported directly to the research and development staff at USAF Headquarters. Meeting participants had to identify a way to delegate technical management down to ARDC while building policy, planning, and control in the DCS/D. Recognizing that they needed to work together, the group participants agreed to a new division of labor and a truce. Perhaps to ensure that both sides held up their ends of the agreement, Putt moved to become the vice commander of ARDC and Lt. Gen. Laurence C. Craigie became the new DCS/D. After Putt's installation in ARDC the skirmishes between the DCS/D and ARDC dissipated and both offices settled into the many critical jobs at hand.[46]

McCormack's report noted the importance of the new development planning process for the DCS/D and the need to delineate clearly the responsibilities of ARDC and the DCS/D. By early 1952 both of those problems had been solved satisfactorily, with the DCS/D concentrating on planning, policy, and top-level control and ARDC focusing on day-to-day management. Planning and policy, however, were not the only major problems faced by the DCS/D and ARDC. McCormack realized that the solutions to those challenges also were tied close-ly to another issue, the problem of managing research versus that of managing large-scale development. To manage both endeavors successfully, McCormack believed that the Air Force needed to implement correctly an idea just then under intense discussion in the Air Staff: the "weapons systems approach."

The Rise of the Weapon System Concept

When McCormack performed his study in the fall of 1951, the weapons sys-tem approach had already been under discussion in the Air Staff for at least six months. Those who espoused this approach recognized that there was a funda-mental difference between the hands-off, loosely controlled methods of manage-ment typical of scientific or technological research and the highly disciplined methods necessary to build a complex weapon to be used in the field. As McCor-mack put it,

> These two objectives are generally different in their demands
> on management, for the first [weapons systems] calls for orga-
> nizational discipline while the second [research] depends for
> success largely on the inventiveness of the technical individual
> and on receptiveness of invention at policy levels, both of
> which often suffer from formal discipline.[47]

Noting that Air Force officers gave "much lip service to their dissimilarity," he found that "in fact, it is difficult to find in our management practices much distinction in method of attack on these essentially dissimilar problem areas. . . ." There were more than three thousand individual projects in the Air Force R&D program and McCormack recognized the impossibility of top management hav-ing the capability or desire to control each one of them. Most of the projects were small component development projects that should be grouped into more coherent packages to be monitored by the director of R&D, while the various laboratories controlled each in detail under the supervision of ARDC. Only those few "major weapons systems" projects deserved top-level attention.[48]

To understand why the distinction between the two was a problem for the Air Force, we must go back in time to understand how the Air Force developed aircraft and the armaments that went into them. The Air Force's organization reflected its prior methods, and McCormack and others believed that those methods were no longer appropriate for the new technologies of the 1950s.

Technology development in the Air Corps and the Army Air Forces before and during World War II focused on developing new aircraft. Typically, the engineering and procurement divisions at Wright Field in Dayton would contract with industry for the aircraft. Wright Field officers and civilians and the operational commands then would test for airworthiness, performance, and mission suitability. Army Ordnance and the Army Signal Corps developed the armaments and electronic gear that Wright Field personnel then integrated into the aircraft. Wright Field workers procured the components and then made modifications to integrate them successfully into the aircraft.[49] Prior to World War II, funding constraints were more important than schedule constraints and that led to a rather leisurely development and testing program commonly described as the "fly-before-you-buy" concept (Figure 5).

After the Air Corps released its specifications, the contractors designed, built, and delivered to the Air Corps a prototype known as the "X-model." The Air Corps

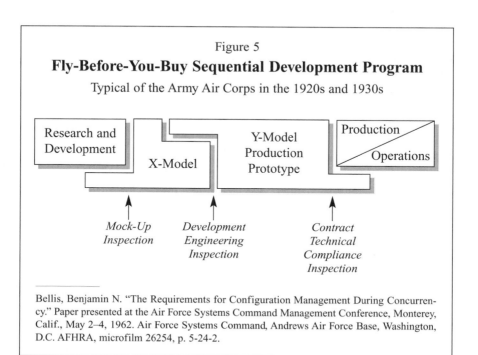

Figure 5

Fly-Before-You-Buy Sequential Development Program

Typical of the Army Air Corps in the 1920s and 1930s

Research and Development

X-Model

Y-Model Production Prototype

Production / Operations

Mock-Up Inspection

Development Engineering Inspection

Contract Technical Compliance Inspection

Bellis, Benjamin N. "The Requirements for Configuration Management During Concurrency." Paper presented at the Air Force Systems Command Management Conference, Monterey, Calif., May 2–4, 1962. Air Force Systems Command, Andrews Air Force Base, Washington, D.C. AFHRA, microfilm 26254, p. 5-24-2.

tested this model and made recommendations for changes to be incorporated into a production model. After completion of X-model testing, the contractor made the recommended design changes and then developed the "Y-model" production prototype. The Air Corps ran another series of tests on the aircraft and made further design recommendations. After approval of the Y-model, the contractor released the production drawings and built the required number of aircraft.[50]

Wright Field personnel managed these projects by assigning to each a project officer with a small supporting staff. In the Bombardment Branch before World War II, Colonel Putt and five other officers managed six aircraft projects with the assistance of a few secretaries. Because of the slow pace of development, the limited role of the government in testing and approving designs, and the fixed-price contracting method typical before the war, the small staff was adequate. Problems of armaments and electronics such as radios became more important as the 1930s drew to a close, and coordination with other organizations became consequently more important and time consuming. Project officers focused on finding weaknesses in the design and making sure that the contractors made the aircraft safe. As Putt noted later, the aircraft contractors and Air Corps pilots often complained that this ruined the performance of their aircraft because it added to aircraft weight.[51]

World War II dramatically changed the situation. With cost constraints removed and high pressure to deliver massive quantities of high-performance aircraft rapidly, Congress, Wright Field personnel, and the Army Air Forces threw the sequential development model out the window. Army Air Forces leaders took virtually every aircraft, whether in production, development, or on the drawing boards, and pressed them forward into production as rapidly as possible. Congress let procurement officers write "letters of intent" to get design and production moving, with the cost negotiations deferred until later. Managers and officers had to balance the need for quantity production with the need for high performance.

The military and its contractors equipped the production lines for runs of large quantities of identical aircraft. If necessary—and it usually was—the Army then ran these aircraft through modification centers where the contractors made the changes necessary for combat. Production lines occasionally would be halted and outfitted for another series of aircraft with the latest modifications, aircraft that went into service immediately. The series–block procedure and the development of modification centers met the need for rapid production, large quantities, and improved performance, albeit at enormous cost.[52]

The Army developed few new aircraft during the war. For two of those new aircraft, the B–29 and the P–61, the Army organized committees to develop the airframe, electronics, and armament together instead of adding them after airframe manufacture and testing.[53] The B–29 committee assembled a consortium of several companies and hundreds of subcontractors to tool up and manufacture B–29s. For that complex and pressurized aircraft, armament and communications were designed together from the start because it was the first aircraft to use computer-controlled fire-control systems integrally connected to the airframe. For these two aircraft, officers considered the entire aircraft as a "system" that included manufacturing and training as well as the hardware. The complexity and organization of the B–29 was a foretaste of things to come.[54]

The B–29 and Manhattan projects were important examples of how to organize to build advanced weapons. The United States also learned from the organization of the German V–2 rocket project. Reporting to General Arnold on German scientific capabilities at the end of the Second World War, Theodore von Kármán noted that one of the major factors in the success of the German V–2 project was its organization:

> Leadership in the development of these new weapons of the future can be assured only by uniting experts in aerodynamics, structural design, electronics, servomechanisms, gyros, control devices, propulsion, and warhead under one leadership, and providing them with facilities for laboratory and model shop production in their specialties and with facilities for field tests. Such a center must be adequately supported by the highest ranking military and civilian leadership and must be adequately financed, including the support of related work on special aspects of various problems at other laboratories and the support of special industrial developments. It seems to us that this is the lesson to be learned from the activities of the German Peenemünde group.[55]

In the Ridenour Report of 1949, the Scientific Advisory Board remembered the lessons of the Manhattan and V–2 projects for organizing large, new technologies. They noted that new systems were far more complex than their prewar counterparts, making it necessary for some engineers to concentrate on the entire system instead of its components only. Project officers also needed greater authority to lead a "task force" of "systems and components specialists organized on a semi-permanent basis." Because the Air Force had few qualified technical officers, the committee recommended that the Air Force draw on the "very important reservoir of talent available for systems planning in the engineering design staffs of the industries of the country."[56] Another year passed before the Air Force acted.

Despite these examples, personnel at Wright Field continued to organize projects along functional lines mirroring academic disciplines and coordinated projects through committees. As Wright Field's Col. Marvin C. Demler put it,

> due to the complexity of the mechanisms which we develop, and our organization by hardware specialties, a very high degree of cooperation and coordination is required between

organizations at all levels. In fact, an experienced officer or civilian engineer coming to Wright Field for the first time simply cannot be effective for perhaps six months to one year while he learns "the ropes" of coordination with other offices. The communication between individuals necessary for the solution of our problems of coordination defy formal organizational lines. . . . [57]

This informal structure was not to continue for much longer. When the Korean War broke out in late 1950, the Air Force found itself with numerous unusable aircraft. In January 1951 Vice Chief of Staff Nathan Twining instructed DCS/D Saville to investigate the Air Force's organization to determine its effect, if any, on the poor aircraft readiness. Saville in turn ordered the formation of a study group, which included Colonel Schriever, to investigate the problem. It was not surprising that the group returned to the Ridenour Report's comments about the lack of technical capability in the Air Force and the problems caused by separating airframe development from component development.[58]

The group completed its study in April 1951 and released an influential staff paper called "Combat Ready Aircraft." Many of the Air Force's later organizational changes came from that paper. The study pinpointed two major problems with current aircraft: requirements based on short-term factors leading to continuous modifications, and insufficient coordination and direction of all elements of the "complete weapon."[59] The former concern probably arose out of the contemporary concern with improving the DCS/D's development planning and future focus, and the latter concern probably arose from the Ridenour Report, from Getting's experience with radar systems integration, and from the examples of the B–29, the V–2, and the Manhattan Project.

To solve these problems the group recommended that the Air Force create an organization and process with responsibility and authority over the complete weapon by adding "planning, budgeting, programming, and control" to the functions of the responsible Air Force organizations. The responsible organizations were to have complete control over the entire project, with that control enforced through full budget authority.[60] Examples of this kind of organization already existed in the Air Force on the guided missile programs. These weapons differed substantially from piloted aircraft, and the separate procurement of airframe, engines, and armament (payload) made little sense.[61] The study group suggested that the Air Force let prime contracts to a single contractor to integrate the entire weapon and that the Air Force organize on a project basis as well.

Changes to the procurement cycle also had to be addressed. The group noted that in World War II decisions to produce aircraft occurred haphazardly and that aircraft rolled off the assembly line directly to the combat units at the same time that they were delivered to testing. Because production continued rapidly and little testing had occurred, invariably the operational and testing units found numerous problems and grounded aircraft for modifications. Contrasting the "fly-before-you-buy" method used before World War II with these World War II "concurrent" methods, the group recommended a compromise. Believing the current emergency did not allow for the fly-before-you-buy sequential approach, but also that the delivery of the initial production aircraft to the combat units was far too wasteful, the group recommended eliminating the X- and Y-model aircraft but slowing the initial production line until test organizations found and eliminated design bugs. Only then should production be accelerated. This, along with project-centered organization and simultaneous planning of all components throughout the life cycle of the weapon, defined the "weapon system concept."[62]

When General McCormack completed his report in November 1951, he referred to the ongoing discussions about the systems approach. He recommended that the rearrangement of the DCS/D office continue along the "pattern of major weapons systems" because that would help reduce the number of Air Force projects (then more than three thousand) and would prioritize them, with only the major systems coming to the attention of the Air Staff. Responsibility for the detailed planning and control necessary for each of these complex weapons had to be lodged in ARDC. Implementing the weapons systems approach within ARDC was likely to succeed because Putt soon moved to become the command's vice commander. He also concurrently held the position of commander of Wright AFB Development Center.[63]

Putt immediately went to work campaigning for systems in the center of resistance, the component developers at Wright Field. The new organization removed power from the functional organizations, and placed it on project officers. The project office acted on a systems basis, making compromises between cost, performance, quality, and quantity. Wright Field Commander Donald Putt admonished the chastened component engineers: ". . . somebody has to be captain of the team, and decide what has to be compromised and why. And that responsibility we have placed on the project offices. . . ." He also stated in no uncertain terms who had the authority, telling the component engineers that they needed to be "sure that all the facts have been placed before them [the project office]. At that time, your responsibility ceases."[64]

Without a large number of technical officers, the Air Force handed substantial authority to industry. Under the weapon system concept, the Air Force "purchased *management* of *new* weapon system development and production" (Reed's italics).[65] A primary aim was to reduce the involvement of the government in the design and supply of equipment. One way to do this was through the prime contractor method whereby the government gave funding to one company to manage and integrate the entire system. A second way was the associate contractor method through which the government hired one company to create the specifications and oversee system development and hired others to develop the component hardware. Both methods required that every contractor "accept the Air Force as the monitor of his plans and progress, with the cautionary power of a partner and the final veto power of the customer." The Air Force stated that it could not "escape its own responsibility for systems management simply by assigning larger blocks of design and engineering responsibility to industry." Although the new process gave industry a larger role, Air Force officers would not remain passive in the project offices or as component developers.[66]

In 1952 and 1953 the systems approach spread through ARDC and the DCS/D office. The DCS/D office coordination process spread to include ARDC, as ARDC laboratories contributed to the "technology push" elements of development planning. Based on the development directives established by the director of research and development, ARDC issued design study directives to perform internal and contracted studies to determine the feasibility of new technology developments. Once those studies were completed, the results were sent to the DCS/D. If results were promising, the DCS/D would issue a development directive. In turn, ARDC would issue contracts to perform the development. By the end of 1952 ARDC separated development into eight phases: task initiation, study, experimentation, development, service test, preliminary production, production, and in-service use.[67]

Adoption of the weapons systems approach throughout the Air Force did not go smoothly because of continuing disagreements between the DCS/D and ARDC on the one hand and the DCS/M and AMC on the other concerning who would control the development of the complete weapons system. The key question that divided the fledgling ARDC and its parent AMC was the question of when "development" ended and when "production" began. If "production" was defined to start relatively early in a weapon's life cycle, then AMC would have greater control; if "development" ended relatively late in the cycle, then ARDC would garner greater power. Not surprisingly, AMC leaned toward a definition of production that encompassed earlier phases of the life cycle and ARDC opted for a late-ending development. In reality, because development continued as long as changes to the weapon occurred and because production began the moment the first prototype was built, there was no "objective" definition that tipped the scales one way or another. Under such circumstances only a respected third party with access to authority could resolve the situation. Once again, James Doolittle was brought in.

In April 1951 Doolittle reported that the ARDC definition should hold because development continued through a system's entire life cycle. Consequently, ARDC should control production engineering. Having been prebriefed by Doolittle, Chief of Staff Vandenberg instructed that ARDC report directly to the DCS/D (instead of through AMC as it had been doing), that Doolittle remain the official arbitrator, and that ARDC provide all engineering services.[68]

In the ongoing balancing act between ARDC and AMC, any new procurement concept such as the weapons system approach would have to be negotiated with AMC. By October 1951 a working agreement between the two organizations surfaced in the issuance of Air Force Regulation 20-10, "Weapons Systems

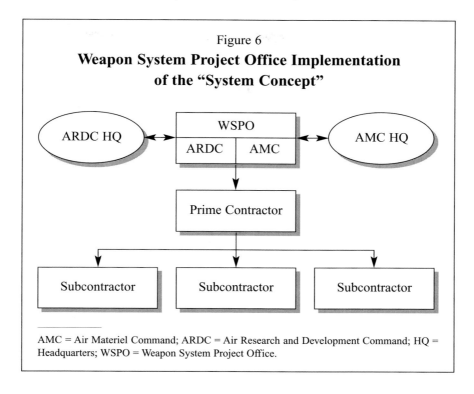

Figure 6
**Weapon System Project Office Implementation
of the "System Concept"**

AMC = Air Materiel Command; ARDC = Air Research and Development Command; HQ = Headquarters; WSPO = Weapon System Project Office.

Project Offices." The regulation specified that every major project should have a Weapons System Project Office (WSPO), with officers from ARDC and AMC in charge (Figure 6).

A marvel of diplomacy, the document stated that during the early portions of development the ARDC representative would be the "team captain," and in the later portions, after a decision to produce the article in quantity, the AMC representative would be the "team captain." In practice, the line between the two was fuzzy and left the two officers to work it out for themselves on the basis of existing circumstances or personalities. The team captain coordinated the activities for the entire project but did not have authority over the other officer. If the two could not agree they would have to take the problem to higher authorities all the way up to the DCS/D and DCS/M at Air Force Headquarters.[69]

The resulting ambiguities continued to cause organizational headaches, leading once again to the intervention of arbitrator Doolittle. This time Doolittle did not feel comfortable forcing a solution so he recommended another Air Staff study to investigate the problem. His only proviso was that the group must protect the importance of R&D. For the time being the Air Staff decided to give the DCS/M, Lt. Gen. Orval Cook, the responsibility for solving the interface problems. In cooperation with DCS/D Craigie, Cook appointed a task group, accordingly known as the "Cook–Craigie Group," to work on the issue. Eventually the group decided that ARDC should keep responsibility for weapon systems until the Air Staff decided to purchase the weapon in quantity and stated their intention in writing.[70]

This decision, formalized in the "Cook–Craigie Procedures" of March 1954 and by modification of Air Force Regulation (AFR) 20-10 in August of that year, ended for the moment the bickering between the development and materiel groups. This might have been partly the result of an unwelcome intrusion by Strategic Air Command (SAC) in early 1953. SAC officers had recommended that a "phasing group" of general officers be assigned for each project to approve engineering changes. ARDC and AMC drew together to defeat the common enemy, vetoing the proposal and stating that this was a prerogative of the ARDC–AMC joint WSPOs. SAC's bid portended events to come but for the time being the two commands had achieved some measure of harmony as they developed complex new weapons.

Conclusion

The Second World War and the Cold War that followed enabled the military to consolidate and extend its relationships with both academia and industry. When in 1947 the Procurement Act gave the Department of Defense the permanent authority to negotiate contracts, military officers enlisted the support of academia and industry to serve military interests. Air Force officers such as Hap Arnold, Donald Putt, and Bernard Schriever promoted and used scientists to create a technologically competent and powerful Air Force. RAND, the SAB, the DCS/D, and ARDC were new means to develop technologies and to tap the resources of research scientists and technologists. Two models for relationships between the Air Force and the scientists evolved. First, RAND, the SAB, and the RDB symbolized the continuing voluntary association of scientists with the military, as had occurred in World War II. However, the DCS/D and ARDC represented new Air Force efforts to gain control over the scientists through a standard Air Force hierarchy. Both models would continue into the future. Through these organizations and their talented members Air Force officers hoped to develop the "Air Force of the future."

It took some time for the Air Force to work out the appropriate roles for their new organizations. Eventually, a corporate-style form took hold with the DCS/D acting as a corporate headquarters that focused on long-range planning, policies, and oversight. ARDC became the field organization devoted to day-to-day management issues. Development planning became a critical new function in the Air Force and Colonel Schriever earned his stripes in that office. RAND Corporation found an appropriate role, eventually creating the ANSER Corporation as a civilian adviser to the Air Staff.

Finally, the complexities of new technologies dictated more complete planning and rigor than had been present in the old Air Force. The "weapons system approach" was the key to integrated planning and control of new weapons, but it had to be worked out in the ongoing struggle between ARDC and Air Materiel Command. After a series of staff studies, again with Schriever as a participant, the "two-headed" Weapon System Project Office was established where AMC and ARDC representatives worked out appropriate roles for themselves. Recognizing that WPSO was a violation of the principle of unity of command, the Air Staff eventually enforced a procedure that led to ARDC leadership until such time as the Air Staff decided to procure the new weapons in quantity and signaled that intention in writing.

In the period from 1945 to 1954, the major organizations through which the Air Force's R&D programs developed were formed. Air Force officers hoped that the organizations would aid their service's future greatly through the development of new technologies. Much to their surprise, by 1953 and 1954 significant new systems began to take shape and they found that these carefully crafted procedures and organizations routinely would be bypassed by dedicated Air Force officers. In the emergency conditions generated by the Soviet test of the hydrogen bomb, as in World War II, existing routines would be tossed out the window and new ones put in their place.

Notes

1. P. Stewart Macaulay, "The Market Place and the Ivory Tower," in Charles D. Flagle, William H. Huggins, and Robert H. Roy, *Operations Research and Systems Engineering* (Baltimore: Johns Hopkins University Press, 1960), p 6.

2. See an extract of the Signal Corps–Wright brothers contract in G. Van Reeth, "Le Contrat avec intéressement comme instrument de gestion," *ESRO/ELDO Bull* 7 (Oct 1969), p 5.

3. Roger E. Bilstein, *Flight in America, From the Wrights to the Astronauts,* rev ed (Baltimore: Johns Hopkins University Press, 1994), pp 70–72; Alex Roland, *Model Research: The NACA 1915–1958* (Washington, D.C.: NASA SP-4103, 1985), pp 73–119.

4. See Irving Brinton Holley, Jr., *Buying Aircraft: Matériel Procurement for the AAF,* vol VII of Stetson Conn, ed, *US Army in World War II* (Washington, D.C.:

Ofc of the Chief of Military Hist, Dept of the Army, 1964), pp 80–86. This is the best introduction to procurement practices prior to and during World War II. For a general history of Army Aviation between wars, see Maurer Maurer, *Aviation in the US Army 1919–1939* (Washington D.C.: Ofc of AF Hist, 1987).

5. Holley, *Buying Aircraft,* pp 89–93, 110.

6. *Ibid.,* pp 113–116.

7. Working paper, Allen Kaufman, *In the Procurement Officer We Trust: Constitutional Norms, AF Procurement and Industrial Organization, 1938–1948* (Cambridge: MIT Defense and Arms Control Studies Prog, Jan 1996), pp 21–29; Holley, *Buying Aircraft,* pp 116–128.

8. A. Hunter Dupree, *Science in the Federal Government: A History of Policies and Activities* (Baltimore: Johns Hopkins University Press, 1986), p 371.

9. Dupree, *Science in the Federal Government,* pp 29–31.

10. David M. Hart, *Forged Consensus: Science, Technology, and Economic Policy in the US, 1921–1953* (Princeton: Princeton University Press, 1998), chapter 5.

11. Holley, *Buying Aircraft,* chapters 13, 15, and 16.

12. Kaufman, *In the Procurement Officer We Trust,* pp 37–42; Holley, *Buying Aircraft,* pp 540–546.

13. See interim rpts, von Kármán to H. H. Arnold, Gen of the AAF, "Where We Stand," Aug 22, 1945, and "Toward New Horizons," Dec 15, 1945, in Michael H. Gorn, ed, *Prophecy Fulfilled: "Toward New Horizons" and Its Legacy* (Washington, D.C.: AF Hist and Museums Prog, 1994).

14. Thomas A. Sturm, *The USAF SAB: Its First Twenty Years 1944–1964* (Washington, D.C.: USAF Historical Div Liaison Ofc, 1967); Donald Ralph Baucom, "AF Images of Research and Development and Their Reflections in Organizational Structure and Management Policies," Ph.D. diss, University of Oklahoma, 1976), pp 11–17, 44–51.

15. John Lonnquest, "The Face of Atlas: Gen Bernard Schriever and the Development of the Atlas Intercontinental Ballistic Missile 1953–1960," Ph.D. diss, Duke University, 1996, pp 57–59; Jacob Neufeld, "Bernard A. Schriever: Challenging the Unknown," in John L. Frisbee, ed, *Makers of the USAF* (Washington, D.C.: Ofc of AF Hist, 1987), pp 281–284; intvw, Bernard A. Schriever with author, Mar 4, 1999, USAF HSO.

16. See Bruce L. R. Smith, *The RAND Corporation: Case Study of a Nonprofit Advisory Corporation* (Cambridge: Harvard University Press, 1966), pp 33–48. Smith noted that one of Arnold's sons married Donald Douglas' daughter in 1943, which he believed reflected the close relationships between the Army Air Corps and the aircraft industry.

17. See Paul N. Edwards, *The Closed World: Computers and the Politics of Discourse in Cold War America* (Cambridge: MIT Press, 1996), chapter 4, for a description of RAND's development and use of ops research and its transformation to systems research analysis.

18. Michael Fortun and Sylvan S. Schweber, "Scientists and the Legacy of World War II: The Case of Operations Research (OR)," *Social Studies of Science* 23 (1993), p 607.

19. See US Cong, *The National Security Act of 1947,* PL 253, 80th Cong, 1st Sess, chapter 343, sec 214, S 758; reprinted in Herman S. Wolk, *Planning and Organizing the Postwar AF, 1943–1947* (Washington, D.C.: Ofc of AF Hist, 1984).

20. For a discussion of the RDB (non)role in missile R&D, see Edmund Beard, *Developing the ICBM: A Study in Bureaucratic Politics* (New York: Columbia University Press, 1976); for a discussion of its utility by the AF RDB representative, see "Historical Documentation, Maj Gen James F. Phillips," Jun 24, 1966, USAF Oral Hist Prog, AFHRA, K239. 0512-787, pp 4–7.

21. Kaufman, *In the Procurement Officer We Trust,* pp 49–55.

22. Harvey M. Sapolsky, *Science and the Navy: The History of the Ofc of Naval Research* (Princeton: Princeton University Press, 1990); Daniel Lee Kleinman, *Politics on the Endless Frontier: Postwar Research Policy in the US* (Durham: Duke University Press, 1995), p 137.

23. Kleinman, *Politics on the Endless Frontier,* p 152.

24. Robert Sigethy, "The AF Organization for Basic Research 1945–1970: A Study in Change," Ph.D. diss, American University, 1980, pp 25–26; Nick A. Komons, *Science and the AF: A History of the AF Office of Scientific Research* (Arlington, Va: Ofc of Aerospace Research, 1966), pp 10–11.

25. For the lab listing, see *Research and Development in the USAF* [Ridenour rpt], SAB, Sep 21, 1949, AFHRA, 168.1511-1, tab B, p 1; on Cambridge, see Sigethy, "The AF Organization," pp 27–29.

26. Memo, E. P. Learned, Special Consultant, to Gen Muir S. Fairchild, Vice CofS, USAF, subj: HQ Organization, USAF, Jul 29, 1949, in Nicholas Roberts, *History of the Directorate of Requirements, Dep CofS; Development,* 1 Jul 1949–30 Jun 1950, AFHRA, K140.01; Memo, Gen Muir S. Fairchild, Vice CofS, USAF, to DCS/P, DCS/O, DCS/M, Comptroller, Inspector Gen, subj: Organizational and Functional Realignment—HQ USAF, Aug 5, 1949, in Roberts, *History of the Directorate of Requirements.*

27. Michael H. Gorn, *Harnessing the Genie: Science and Technology Forecasting for the AF 1944–1986* (Washington, D.C.: Ofc of AF Hist, 1988), p 48.

28. Intvw, Lt Gen Donald L. Putt with James C. Hasdorff, USAF Oral Hist Prog, Apr 1–3, 1974, AFHRA, K239.0512-724, p 79.

29. *Ibid.*, pp 79–82.

30. *Research and Development in the USAF* [Ridenour rpt], SAB, Sep 21, 1949, AFHRA, 168.1511-1; Beard, *Developing the ICBM,* pp 107–117; intvw, Putt, April 1–3, 1974, pp 83–85. For other histories of these events, see Michael H. Gorn, *Vulcan's Forge: The Making of an AF Command for Weapons Acquisition (1950–1985)* (Washington, D.C.: Ofc of Hist, AFSC, 1989), pp 7–19; Sigethy, "AF Organization"; Baucom, "AF Images"; Komons, *Science and the AF;* Dennis J. Stanley and John J. Weaver, *An AF Command for R&D, 1949–1976: The History of ARDC/AFSC* (Washington, D.C.: Ofc of Hist, HQ AFSC, 1976).

31. J. P. Thornton, *History of the DCS/D, 1 Jan to 30 Jun 1951,* AFHRA, K140.01, p 4; "Gen Doolittle's Summ," Jan 15, 1952, MDP, AFHRA, 168.7265-235.

32. For a characterization of Saville, see Kenneth Schaffel, *The Emerging Shield: The AF and the Evolution of Continental Air Defense, 1945–1960* (Washington, D.C.: Ofc of AF Hist, 1991), pp 84–86.

33. Ivan Getting, *All in a Lifetime: Science in the Defense of Democracy* (New York: Vantage, 1989), pp 225–227. Getting also reported that he did not accept the position until Ridenour's situation with respect to ARDC was defined. He believed the two of them would not get along very well.

34. Getting, *All in a Lifetime,* pp 228–230.

35. J. P. Thornton, *History of the DCS/D, 1 Jul to 31 Dec 1950,* AFHRA, K140.01, p 14.

36. David R. Jardini, "Out of the Blue Yonder: The RAND Corporation's Diversification into Social Welfare Research, 1946–1968," Ph.D. diss, Carnegie Mellon University, 1996, pp 56–64.

37. Jardini, "Out of the Blue Yonder," pp 64–68; J. P. Thornton, *History of the DCS/D, Jul 1951 to Jun 1952,* AFHRA, K140.01, p 6.

38. Getting, *All in a Lifetime,* p 240; intvw, Getting, Oct 30, Nov 6, 1998.

39. Getting, *All in a Lifetime,* pp 240–241; intvw, Getting, Oct 20, Nov 6, 1998; Thornton, *History of the DCS/D, Jul 1951 to Jun 1952, pp 3–4;* intvw, Bernard A. Schriever with author, Mar 4, 1999, USAF HSO.

40. See Gorn, *Vulcan's Forge,* pp 29–31. The chart on p 30 of that book shows that the role of the DCS/D in the development cycle is labeled "Development Cycle Prior to ARDC," but that technically is not correct because ARDC and the Ofc of the DCS/D came into being at the same time. It is true, however, that in 1950 and early 1951 the procedures were developing, and ARDC did not have the expertise to play a significant role in the process.

41. Getting, *All in a Lifetime,* p 241.

42. Intvw, Putt, Apr 1–3, 1974, pp 44–45; intvw, Gen Bernard A. Schriever with Dr. Edgar F. Puryear, Jr., USAF Oral Hist Prog, Jun 15 and 29, 1977, AFHRA, K239.0512-1492, pp 5–6; Lonnquest, "Face of Atlas," pp 57–64; intvw, Bernard A. Schriever with author, Mar 4, 25, and Apr 13, 1999.

43. Thornton, *History of the DCS/D, 1 Jul to 31 Dec 1950,* pp 9–11.

44. Ltr, Brig Gen D. N. Yates, Acting DCS/D, to Brig Gen James McCormack, Jr., Aug 24, 1951, in Lt Col Robert A. Maddocks and Lt Col John L. Gregory, Jr., *History of the Directorate of Research and Development Ofc, DCS/D, HQ USAF, 1 Jul 1951 to 31 Dec 1951,* AFHRA, K140.01; Thomas P. Hughes, *Rescuing Prometheus* (New York: Pantheon, 1998), p 63.

45. Memo, Brig Gen James McCormack, Jr., Special Asst to the DCS/D, to Gen Putt, subj: Rpt on AF R&D Organization, with Particular Reference to DCS/D–ARDC Relationships, Nov 26, 1951, in Maddocks and Gregory, *History of the Directorate,* pp 5–6; Thornton, *History of the DCS/D, Jul 1951 to Jun 1952,* p 8.

46. Maddocks and Gregory, *History of the Directorate,* pp 1–3; Thornton, *History of the DCS/D, Jul 1951 to Jun 1952,* pp 5–7.

47. Memo, McCormack to Putt, Nov 26, 1951, in Maddocks and Gregory, *History of the Directorate,* p 4.

48. *Ibid.,* pp 4–7.

49. "Gen Doolittle's Summ," MDP.

50. Presentation, Benjamin N. Bellis, L/Col USAF Ofc DCS/S, "The Requirements for Configuration Management During Concurrency," at AFSC Management Conference, Monterey, Calif, May 2–4, 1962, AFSC, Andrews AFB, Washington, D.C. AFHRA, microfilm 26254, pp 2-24-1–4; AF study prepared by DCS/D, "Combat Ready Aircraft," Apr 1951, MDP, AFHRA, 168.7265-236, pp 16–18.

51. Intvw, Putt, Apr 1–3, 1974, pp 23–24, 123.

52. Holley, *Buying Aircraft,* chapter 20; DCS/D, "Combat Ready Aircraft," pp 17–18.

53. Lonnquest, "Face of Atlas," pp 213–214.

54. Jacob Vander Muelen, *Building the B–29* (Washington, D.C.: Smithsonian Institution Press, 1995), pp 11–29.

55. Von Kármán, "Where We Stand," p 37.

56. *Research and Development in the USAF* [Ridenour rpt], pp IX-1-2.

57. Presentation, Col M. C. Demler to CofS and Comptroller, WADC, Management Practices, Feb 15, 1952, AFHRA, 168.7265-235, p 4.

58. DCS/D, "Combat Ready Aircraft," p 2.

59. *Ibid.,* pp 9–14.

60. *Ibid.,* p 15.

61. *Research and Development in the USAF* [Ridenour rpt], p IX-2.

62. DCS/D, "Combat Ready Aircraft," pp 19–24.

63. Memo, McCormack to Putt, Nov 26, 1951, in Maddocks and Gregory, *History of the Directorate,* pp 4–7, 12–14; see also Gorn, *Vulcan's Forge,* chart 3, p 24.

64. Presentation, Maj Gen Donald L. Putt, Commanding Gen, WADC, to Chiefs of WADC Labs, "Organizational Philosophy of the AF," Apr 2, 1952, AFHRA, 168.7265-235, pp 6–8.

65. Robert J. Reed, "New AF Policy Means More Competition—More Selling," *Aviation Age* 19 (Aug 1953), p 21.

66. *Ibid.,* p 22.

67. See Gorn, *Vulcan's Forge,* p 29.

68. *Ibid.,* pp 19–20.

69. Air Force Regulation 20-10, "Weapon System Project Office," Oct 16, 1951; see also Gorn, *Vulcan's Forge,* pp 28–29.

70. The Cook–Craigie group included Harvard Business School professor Edmund Learned (who previously had done organizational studies for HQ USAF), Bell Labs chief Mervin Kelly, Brig Gen Floyd Wood of ARDC, Edward Wells of Boeing, and Col William E. Sault of AMC. See Gorn, *Vulcan's Forge,* p 27.

Chapter 3

Building the Weapon of the Future: Intercontinental Ballistic Missiles

> We have found that concurrency is as unforgiving to inept management principles as a high performance aircraft is to pilot error. In fact, it requires MORE formality, not LESS.
>
> Lt. Col. Benjamin Bellis, 1962[1]

While officers in the office of the deputy chief of staff, development at Air Force Headquarters learned to focus on executive management functions such as development planning and policy, officers in Air Research and Development Command took over the day-to-day management of new technologies. With projects ranging from small research and study contracts to component and large-scale systems development, ARDC officers had their hands full. Although the bulk of the Air Force's projects understandably concentrated on aircraft and aircraft components, missile development had become an important and potentially troublesome competitor for resources in the pilot-oriented culture of the Air Force.

Missiles, and particularly ballistic missiles, were disruptive to the Air Force's culture, operations, and organization in several important ways. First, and most obvious, missiles had no pilots so they relegated humans simply to getting the missile somewhere within range of the target and then pushing a button. Second, because missiles never returned after being fired, maintenance and long-term operations amounted to storage and occasional refurbishment instead of the ongoing repairs typical for aircraft. Third, missile development and production differed significantly from aircraft. The concepts of the X-model and the Y-model aircraft prototypes simply had no meaning for missiles, and the "fly-before-you-buy" concept that permitted testing before full-scale production no longer applied. Unlike aircraft, for which a few prototypes could be built and tested with dozens or hundreds of flights, missiles were one-time-use products and testing required a full production line. Missile up-front costs were proportionally higher than those for aircraft. Finally, as discussed earlier, missiles involved a whole slew of new technical issues to be overcome. Simply put, many of the Air Force's organizational and technical processes did not work for missiles.

By the early 1950s the differences between missiles and aircraft were becoming apparent to some Air Force officers. When the ballistic missile became a feasible weapon with the predicted size reduction and huge yield of fusion (hydrogen) reactions, nuclear scientists and Air Force officers realized that the missile likely would become the ultimate strategic weapon. The Eisenhower administration and the Air Force decided to press intercontinental ballistic missiles into full-scale "crash" development. This decision, when combined with the technical implications of ICBM development, testing, and operations, yielded a new set of organizations and organizational processes. Borrowing from then-current ideas about applying the systems approach and the existing division of labor between ARDC and AMC through project offices, recently promoted Brig. Gen. Bernard Schriever led his new organizations to develop innovative processes tailored to ICBM development.

Schriever insisted on having sufficient authority to carry out his mission and that led to an extraordinary degree of independence from the Air Force's normal processes, procedures, and oversight. In their relatively isolated "organizational space" Schriever and his fellow officers and organizations devised processes that became remarkably influential. The scope of his Western Development Division (WDD) and its successor, the Ballistic Missile Division (BMD), expanded from developing missiles and their reentry vehicles to developing spacecraft as well. In addition, the perceived success of these new approaches influenced Air Force leaders to adopt these methods for other systems, eventually including aircraft and command and control systems. Schriever's organization became the model for USAF large-scale technology development.

ICBMs and Formation of the WDD

Ballistic missile programs had languished at a low priority during and after World War II while the Air Force concentrated its efforts first on manned bombers and then on jet fighters for the Korean War. Air Materiel Command ran the Air Force's program and monitored the competing Army and Navy missile programs. The Research and Development Board tried without success to coordinate the various programs after 1947.[2]

The rapidly escalating Cold War provided the impetus to transform the loosely organized missile projects. The USSR's successful testing of the atomic bomb in 1949 and a hydrogen bomb in 1953 led to a major shift in the pace and priority of missile development in the United States.[3] As the chief of the Development Planning Office in early 1953, Schriever learned of the success of U.S.

James Doolittle, renowned World War II bomber pilot, was a strong proponent of the need for a research and development organization.

thermonuclear tests from the Scientific Advisory Board. He immediately recognized the implications of this news for ICBMs, and within days he met with renowned mathematician John von Neumann at his Princeton office. Von Neumann confirmed Schriever's opinion that scientists would soon develop nuclear warheads of small enough size and large enough explosive power to be placed on ICBMs. Because of the missiles' speed and invulnerability, they were the preferred method for nuclear weapons delivery if such delivery could be achieved. Schriever realized that he needed official evidence so he talked with James Doolittle, who approached Chief of Staff Vandenberg to get the SAB to investigate the question.[4]

The SAB's Nuclear Weapons Panel, headed by von Neumann, reported to the Air Staff in October 1953. In the meantime, Trevor Gardner, assistant to the secretary of the Air Force, volunteered to head a DOD study group on guided missiles. Gardner learned of the progress of the Atlas program from its contractor, General Dynamics Convair Division, and met Dr. Simon Ramo, the head of Hughes Aircraft Company's successful air-to-air missile program, the Falcon. On the basis of the results of Gardner's study group, Gardner and Air Force Secretary Harold E. Talbott formed the Strategic Missiles Evaluation Committee, or

Maj. Gen. Bernard A. Schriever is congratulated on his promotion to two-star status by Simon Ramo (right), president of the Ramo-Wooldridge Corporation, and Trevor Gardner, special assistant to the Air Force for Research and Development. Gardner knew Ramo from their days at General Electric in the 1940s—a key factor in the selection of Ramo-Wooldridge to work with the Teapot Committee.

"Teapot Committee," to investigate and recommend a course of action for strategic ballistic missiles.[5]

Von Neumann headed the committee and Gardner selected Simon Ramo's newly created Ramo-Wooldridge Corporation to do the paperwork and manage the day-to-day operations of the study.[6] Gardner knew Ramo well for they had lived in the same apartment building years earlier when both worked at General Electric. Ramo eventually took a job at Hughes Aircraft, assembling one of the nation's largest concentrations in industry of scientists and engineers with graduate training. This unusual abundance of highly trained technical personnel at Hughes concerned Secretary of Defense Charles Wilson, who called Ramo into his office in the early 1950s. Under Ramo, Hughes Aircraft had created a dominant position in supplying the electronics in many military fighters and missiles. Wilson was worried about the eccentricities of company owner Howard Hughes and he urged Ramo to form a separate company to break up the supply monopoly. Ramo and his colleague Dean Wooldridge did just that in 1953, with Thompson Products as a major investor in the new company. When Gardner and the Air Force came calling, the new company was still small and looking for business.[7]

In February 1954 the Teapot Committee recommended that ICBMs be developed "to the maximum extent that technology would allow." They also recommended creating an organization that harkened back to the Manhattan Project and MIT's Radiation Laboratory of World War II:

> The nature of the task for this new agency requires that overall technical direction be in the hands of an unusually competent group of scientists and engineers capable of making systems analyses, supervising the research phases, and completely controlling the experimental and hardware phases of the program—the present ones as well as the subsequent ones that will have to be initiated.[8]

On May 14, 1954, the Air Force made Atlas its highest R&D priority. Because Convair in San Diego had the most advanced ICBM program and because the majority of the aircraft industry was located in southern California, the Air Force established their new IBCM development organization in a vacant church building in Inglewood, near the Los Angeles airport. The Air Force initially planned to give the job of commander to Maj. Gen. James McCormack, who had earlier analyzed the problems of R&D management as a special assistant to the DCS/D, but health problems forced him out of the picture. Air Force leaders placed the recently promoted General Schriever in command of the new Western

The Western Development Division found early temporary quarters in what had been St. John's Catholic School in Inglewood, California. Pictured here is the site in 1954. The site's casual appearance lent itself to the informality of Schriever's management approach.

Development Division of ARDC and he took command on August 2, 1954. Because the Teapot Committee had recommended creating a "Manhattan-like" project organization, one of Schriever's first tasks was to see if that recommendation made sense and, if so, to determine who would oversee the technical aspects of the project.[9]

After comparing the technical features of atomic weapons and ICBMs, Schriever concluded that the missiles were significantly more complex because of the heterogeneity and immaturity of the technologies in their component parts. That conclusion made the Manhattan Project's organization a model with only limited utility for ICBMs. Because neither he nor the scientists believed that the Air Force had the technical expertise to manage the ICBM program, Schriever had two options: he could hire Convair as prime contractor or he could hire Ramo-Wooldridge as the system integrator and Convair and other contractors as associate contractors. The Air Force had used the prime contractor procedure on earlier weapon system programs but the procedure gave the Air Force relatively little leverage. It also assumed that the prime contractor had the wherewithal to design and build the product. Under that method, the Air Force funded the prime contractor, who then subcontracted for other components. Schriever was already unhappy with Convair's lack of subcontracting on Atlas. He believed that to the program's detriment Convair was keeping "in-house" such elements as guidance and electronics—components with which Convair had little experience.[10]

Schriever also was deeply influenced by his scientific advisers with whom he had worked for nearly a decade. Von Neumann and his fellow scientists on the Nuclear Weapons Panel of the Scientific Advisory Board believed the Soviet threat to be just as serious as the Nazi threat had been. They held that this threat required a response just as extraordinary as the Manhattan Project a decade earlier, one that brought together the nation's best scientists to marry ballistic missiles to thermonuclear warheads. Because the government could not hire the needed expertise on civil service pay scales, Schriever and the Nuclear Weapons Panel believed a civilian organization would be necessary to mobilize scientific expertise. As Schriever later explained,

> Complex requirements of the ICBM and the predominant role of systems engineering in insuring that the requirements were met, demanded an across-the-board competence in the physical sciences not to be found in existing organizations. Scientists rated the aircraft industry relatively weak in this phase of engineering, which was closely tied to recent advances in physics. The aircraft industry, moreover, was heavily committed on major projects, as shown by existing backlogs. Its ability to hire the necessary scientific and engineering talent at existing pay-scales was doubted, and with the profit motive

dominant, scientists would not be particularly attracted to the low-level positions accorded to such personnel in industry.[11]

This was one of the clearest statements of Schriever's belief in the scientific ethos and its postwar alliance with the military. He believed that systems engineering was tied closely to advances in physics and that only generically trained physical scientists had the "across-the-board competence" required for systems engineering. In Schriever's opinion, engineers in the aircraft industry, who generally had little or no experience in new nuclear weapons, rocketry, and electronics technologies, could not lead the effort.

Through many years of interaction with scientists, Schriever had developed a deep admiration for them and for their views:

> I became really a disciple of the scientists who were working with us in the Pentagon, the RAND Corporation also, so that I felt very strongly that the scientists had a broader view and had more capabilities. We needed engineers, that's for sure, but engineers were trained more in a, let's say a narrow track having to do with materials than with vision.[12]

To capitalize on the vision and expertise of physical scientists and mathematicians such as von Neumann and von Kármán, Schriever created an organizational scheme whereby the leading scientists could play a major role in guiding the ICBM program. Schriever believed that scientists, inspired by the quest for knowledge instead of profits, needed a purer and stronger organizational position than they would have in the aircraft industry. Consequently, he opted for an arrangement that placed scientists in a powerful position in which they directed industry without becoming low-paid civil servants or being contaminated by capitalist motives.

Following the recommendations of von Neumann's committee, Schriever selected the associate contractor method and hired Ramo-Wooldridge Corporation (R-W) for systems engineering and integration.[13] Being free of civil service regulations and without significant contracts or conflicts of interest, R-W could hire the requisite scientific and technical talent. The Air Force maintained more leverage because R-W had few other contracts and no production capability and the Air Force could direct the firm more easily. Donald Quarles, the deputy secretary of defense for research and development, insisted that R-W be placed with "line" responsibility for systems engineering and integration. This meant that they not only advised the Air Force but had directive authority of their own. Quarles's background with AT&T, where the Bell Telephone Laboratories had that kind of authority over other parts of AT&T and other organizations, made him comfortable with such an arrangement. The aircraft industry disputed this unusual arrangement, fearing that it established a precedent for "strong systems manage-

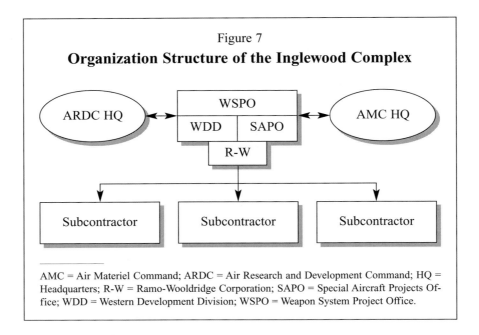

Figure 7

Organization Structure of the Inglewood Complex

AMC = Air Materiel Command; ARDC = Air Research and Development Command; HQ = Headquarters; R-W = Ramo-Wooldridge Corporation; SAPO = Special Aircraft Projects Office; WDD = Western Development Division; WSPO = Weapon System Project Office.

ment control" by the Air Force and that it might create a powerful new competitor with inside information on Air Force contracts and contractor capabilities.[14]

Schriever had the luxury of handpicking his officers from throughout the Air Force, and he took liberal advantage of his authority by selecting the best and brightest officers from ARDC. His talented staff quickly took charge of ICBM development. Because AMC retained authority for procurement it set up a field office alongside Schriever's ARDC staff in Inglewood. By early September, when Air Force Headquarters signed off on the selection of Ramo-Wooldridge, the triumvirate of organizations—the WDD, the Special Aircraft Projects Office (SAPO) of AMC, and R-W—completed the government's project organization.[15] Figure 7 shows the structure of Schriever's new organization.

Scientific influence in the Air Force reached its height with Schriever's rise to head the ICBM programs, his use of the Teapot Committee recommendations, and the hiring of Ramo-Wooldridge to bring together the nation's scientists in support of the ICBM program. The alliance of the WDD and R-W was to take advantage of the larger alliance of military officers and scientists, with the officers supplying the funding and objectives and the scientists providing credibility and technical direction. However, Schriever soon found that scientific expertise alone would not suffice. His next battle would be to establish the authority and credibility of his team in the face of skepticism at Air Force Headquarters and the outright hostility of the aircraft industry.

Establishing WDD's Authority

With his organizational foundations set, Schriever's immediate task was to push ICBM development forward as rapidly as possible, and within a year to create a detailed plan for that effort. Headquarters had taken its first steps with the appointment of an assistant chief of staff for guided missiles in April and the creation of a "skeleton" general operational requirement in August. Its control and oversight would come through the budget process so Schriever knew he had to keep a low budget profile. The budget was to be reprogrammed (reallocated) from several Air Force organizations so he had to be careful not to ask for too

A meeting of the Scientific Advisory Committee (OSD) in the fall of 1955—the first such meeting held at the Air Force Ballistic Missile Division headquarters in Englewood, California. The committee was charged with investigating and recommending a course of action for ballistic missiles. Seated at the table facing Maj. Gen. Bernard Schriever at the speaker's stand are George McRae of Sandia Corporation, Charles Lindbergh, Gen. Thomas Power, Assistant Air Force Secretary Trevor Gardner, committee chairman Dr. John von Neumann, Col. Harold Morton, H. Guyford Stever of MIT, Dr. Clark Millikan of Caltech, and Gen. Charles McCorkle. Others in the group include General Canterbury, Dr. Charles Lauritsen of Caltech, Dr. George B. Kistiakowsky of Harvard University, Dr. Jerome Wiesner of MIT, and Louis Dunn of R-W.

much right at the start. He set several task teams to work figuring out where to get the money within this constraint. Schriever knew that the massive budget he would need over the long haul would require additional congressional appropriations and that he would have to defend his plan of action and its costs vigorously and effectively. To put off that day of reckoning, in October 1954 he requested a relatively small budget, anticipating that there would have to be a major readjustment in the spring. Already he realized that "this support can be obtained by carefully planned and formalized action at the highest levels in the administration." In this breathing space he would have to develop his technical plans, costs, and justifications and a shrewd political strategy to advance his agenda.[16]

With the Atlas already chosen as the vehicle to develop, Schriever's team had to select the contractors who would build it. The missile's design was not yet settled so contractor selection would have to be done on the basis of company capabilities instead of particular submitted designs. Bypassing standard procurement regulations and procedures, Schriever ordered Ramo-Wooldridge to let subcontracts to potential suppliers to get them involved and educated on the program. This allowed R-W to assess the contractors as well as speed the development and procurement processes. Schriever instructed his team to develop a screening procedure to determine who should and should not be funded and assessed. On such a large program, Schriever also warned his SAPO (AMC) procurement specialist, Col. William Sheppard, that this would not work for contractors "sent to us from top-side"—that is, from Schriever's superiors. Recommendations from Headquarters or other politically sensitive individuals had to be treated gingerly to avoid alienating potential political support. Schriever could not apply his screening procedure to those recommended organizations. In the meantime, Schriever could not ignore all of the Air Force's procurement procedures. He had his team begin rapid development of performance specifications and perform "pre-bidding activities" as expeditiously as possible to prepare for a competitive bidders' conference. Because of its in-depth knowledge gained through these contracts, Schriever had R-W contribute to the Source Selection Boards, at least by providing inputs as requested by the Air Force. This was a serious (and possibly illegal) departure from standard procurement policy, which required that only government officials control contractor selection.[17]

In accordance with the findings of the Teapot Committee and his own assessment of the situation, Schriever directed R-W and his Air Force team to reassess the design of Atlas and to determine the proper role for Convair. Convair had been working on the design and development of Atlas since January 1946 and management understandably believed that the company's efforts should have been rewarded with a prime contract to build, integrate, and test the vehicle. They were particularly upset because the Air Force had canceled the program in 1948. Convair believed in the project's promise and poured their own money into it until Air Force interest revived. Schriever's selection of R-W angered them and they vigorously campaigned against Schriever and the upstart company.

Although the Teapot Committee and Schriever were unimpressed with Convair's overall capability to build the Atlas, most aircraft industry leaders believed themselves capable of building ICBMs and backed Convair against the Air Force's new handmaid, R-W. The selection of R-W to be the "systems engineering and technical direction" contractor violated the Air Force's own professed adherence to the systems approach. Through its regulations the Air Force stipulated that WSPOs staffed by ARDC and AMC representatives would let contracts to prime contractors, who in turn would ensure that the entire system was properly designed and produced. Prime contractors did not want their authority usurped by upstart R-W any more than did Convair. Even more important, they predicted the loss of corporate profits because of stronger Air Force technical cognizance, and the emergence of R-W as a dangerous new competitor. As we shall see, their fears were justified on both counts.[18]

By December 1954 Convair's anti–R-W campaign was in full swing. As a former employee of Douglas Corporation, one of the leading aircraft manufacturers, RAND Corporation president Frank Collbohm believed that Convair was quite capable of building Atlas and that the missile was ready for production engineering. He broadcast his opinion from his position at RAND and this caused the Air Force some embarrassment. Through the Aircraft Industries Association (AIA) Convair also recruited the other aircraft manufacturers to consider a strong protest to the Pentagon. Schriever recognized the danger, stating that "the project office has been no match for the powerful pressure that industry can, and has, exerted at political and high military levels." The Air Force's reliance on industry, usually a political asset, was turning into a liability in this case.[19]

Convair and the AIA leveled a number of allegations at Ramo-Wooldridge and the Air Force's approach. Schriever answered them in a letter to Lt. Gen. Thomas Power, the commander of ARDC, in the winter of 1954–55. First, the AIA and Convair alleged that Simon Ramo and Dean Wooldridge had been unethical in leaving Hughes to form their own company. Schriever considered that ridiculous because virtually every aircraft company had started the same way and had received government support through contracts in precisely the same manner. Along with Collbohm, the AIA and Convair claimed that the Atlas was ready to go to production engineering and that Schriever and his coterie of scientists were delaying the program needlessly through their endless studies and analyses. They also claimed that inserting R-W into the process reduced competition. Schriever retorted that the associate contractor approach opened up competition to electronics, computer, and inertial guidance manufacturers because they could compete directly instead of through the prime contractors.[20]

Schriever's organization and Convair sparred for the next few months before Convair resigned itself to R-W's presence. Convair realized that they needed to hire highly educated scientists and engineers to appease the Air Force's scientific advisers and to gain electronics capability. They advertised their hiring successes through a newsmedia blitz. Schriever groaned that this "publicity campaign is strictly for the birds" and directed one of his subordinates to evaluate the media stories "with the objective of exposing them for what they really are." On the Air Force's part, Schriever had to prove Convair wrong and he recognized that he needed to place some restrictions on R-W to maintain any semblance of support from the aircraft industry. Accordingly, in a memo dated February 24, 1955, the Air Force prohibited R-W from engaging in hardware production.[21]

That prohibition, however, did not end the difficulties among Convair, R-W, and the Air Force. Louis Dunn,[22] R-W's Guided Missile Research Division manager, severely criticized Convair's performance in March 1955. As Dunn explained it to Ramo, Convair had fine young engineers but it lacked technically competent managers. The firm did not address problems with sufficient speed or vigor to match the urgency of the situation. Finally, Convair's continuing political, sales, and propaganda drives irritated Dunn, along with their continuing arguments with R-W and the Air Force. Dunn's deputy, Rube Mettler, went to a conference at Convair on the first of April and he found that Convair continued to have difficulties accepting R-W's role. This finding led to a series of meetings

Lt. Gen. Bernard A. Schriever, commander of Air Research and Development Command (center left), greets Dr. Louis Dunn, R-W's Guided Missile Research Division manager. Also pictured (left to right) are Brig. Gen. Charles Terhune, Schriever's technical director; Brig. Gen. Don Coupland; Dean Wooldridge and Simon Ramo, R-W principals; and Rube Mettler, Dunn's deputy.

between Ramo and Convair president, Gen. Joseph McNarney (retired), intended to resolve the tensions. Finally, in June 1955 Schriever himself met with McNarney. After those meetings, Convair's leaders realized that they had to quit hindering the Atlas program but, as we shall see, they continued to oppose R-W through other channels.[23]

To battle Convair's campaign against R-W and the WDD, Schriever had to show that Convair and the aircraft industry were not as competent as they believed themselves to be. Some of that evidence came from service personnel like Air Force plant representative Col. James McCarthy, who sent a letter to Schriever in April 1955 stating that Convair deliberately misled his officers and remained adamant about being the dominant contractor on the project. Like Louis Dunn, McCarthy complained about Convair's weak technical leadership. Soon Schriever would have substantial evidence of R-W's competence and Convair's shortcomings.[24]

Schriever initially assigned R-W three major tasks: establish and operate the facilities for the Inglewood complex, assess contractor capabilities, and investigate the design of Atlas. R-W made its first important contribution in the design task. The required size and performance of the missile depended largely on the size of the payload—that is, the warhead and the reentry vehicle. Small changes in the size of the payload led to large changes in the required size of the launch vehicle. Working with the Atomic Energy Commission and other scientists, R-W scientists and engineers found that Convair's nose cone design yielded such high heat on reentry that they could not perform laboratory experiments to test the design. Conversely, a blunt nose cone design decreased the temperatures sufficiently to permit laboratory testing. Furthermore, by performing a number of trade studies (studies that compare the effect of one variable, such as warhead weight, with another, such as the resultant target damage) R-W personnel found that the new blunt nose cone design decreased the weight by half, from about 7,000 to 3,500 pounds. That reduction in turn decreased the required size of the launch vehicle from 460,000 to 240,000 pounds, and reduced the number of engines from five to three. Such dramatic improvements in the Atlas design discredited much of Convair's argument of expertise and capability and convinced Schriever, his team, and his superior officers that the selection of R-W had been correct.[25]

Despite his ongoing battle with Convair, Schriever realized he could not do without them. In October 1954 he recommended that Convair continue as the airframe and integration contractor for Atlas, and in January 1955 he awarded them a letter contract. Similarly, the engine contractor, North American Aviation (NAA), had to be put under contract immediately to get the ICBM program moving. NAA received its letter contract in October 1954 as well. Finally, the nose cone design had to be tested so the Special Aircraft Projects Office placed Lockheed under contract in January 1955 to build the X–17 reentry test vehicle.[26]

The most significant technical issue facing Schriever's group in the fall and winter of 1954 was the high degree of uncertainty surrounding virtually all

The two-stage Titan I missile was liquid propelled. In its first test in December 1960, both the test vehicle and its test facilities at Vandenberg Air Force Base blew up during propellant loading.

elements of the design. At this early stage of missile development, they simply could not predict which elements of the design would work and which might not. Ramo-Wooldridge had been investigating a two-stage vehicle and the initial results looked promising. By October 1954 Schriever had recommended to ARDC Commander General Power that a two-stage vehicle be developed as a backup design to the Atlas. By March 1955 Power was convinced and he formally proposed to Air Force Headquarters that an alternate-configuration ICBM be funded. Secretary of the Air Force Harold Talbott approved the proposal at the end of April, and by May 1955 WDD was working on Atlas, the two-stage Titan, and a tactical ballistic missile (ultimately known as the Thor).[27]

With the selection of contractors for Atlas and the essential configuration of the missile settled, Schriever's group prepared to do battle with the budget masters. This required going through the Air Force Headquarters DCS/D, and its series of documents—the development planning objectives, the general operational requirements, and the development directives. Only with those hurdles completed would the official budget be allocated—programmed—for ICBM development.

Schriever and Colonel Sheppard, his deputy commander for program management, debated how best to fund the program. One possibility was to allocate the funds into a number of different budgets and then pull them back together in the SAPO. That approach was desirable because it effectively hid the true budget amounts from effective oversight, but Sheppard didn't "know how one would go about hiding such a large element in an Appropriations Act." With programmatic invisibility unlikely, the best approach was to have a "separately justified and separately managed lump sum. . . ."[28]

Schriever already had discussed this approach with assistant secretary of the Air Force Trevor Gardner and the two of them began to plot their political strategy to accelerate the program. Although Schriever could direct the activities of his own staff, many of his actions, particularly those associated with the budget, required coordination with and justification to various organizations. In fact, Schriever found that planning and performing the ICBM program required him to coordinate with numerous organizations and deal with five separate budget appropriations. Frustrated with the delays, Gardner and Schriever decided that they had to increase Schriever's authority and funding and decrease the number of organizations that could oversee and delay ICBM development. In the meantime, Schriever's crew prepared their plans and presented them to the staff at the office of the DCS/D. That presentation led to the issuance of the general operational requirements and development directives in July 1955.

Schriever and Gardner recognized that they needed political support, but also that ICBMs were so novel that they needed to prepare a simple and clear presentation describing ICBMs, their importance to air defenses, and the resources needed to produce them. Schriever went so far as to consider "an animated cartoon type of approach . . . following the 'Disneyland' pattern." Finding support in Congress and within the Eisenhower administration, Gardner and Schriever briefed the president in July 1955, and with John von Neumann's timely support they eventually convinced him and Vice President Nixon to make ICBMs the nation's top defense priority.[29] Figure 8 illustrates the organizational structure that the Gillette and Robertson committees investigated.

With the president's endorsement in hand by September 1955, Schriever presented to Gardner the entire Air Force approval process, which required thirty-eight Air Force and DOD approvals or concurrences. The challenge was most significant for the development of facilities for ICBM testing. Appalled at the bureaucracy, Gardner had Schriever show the process to Secretary of the Air Force Donald Quarles, who tasked them to come up with recommendations to reduce the paperwork and delays. Following what was by now standard procedure, Gardner and Schriever formed a study group that, as Schriever put it later, ". . . we loaded . . . pretty much with people who knew and who would come up with the right answers. . . ." Hyde Gillette, the deputy for budget and program management in the Office of the Secretary of Defense (OSD), chaired the group charged with recommending changes to streamline ballistic missile management to speed development. Secretary of Defense Charles Wilson took interest in the subject and appointed a second committee headed by Deputy Secretary of Defense Reuben B. Robertson at the DOD level to perform a concurrent investigation.[30]

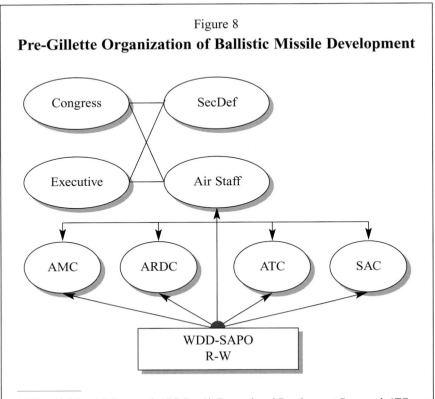

Figure 8

Pre-Gillette Organization of Ballistic Missile Development

AMC = Air Materiel Command; ARDC = Air Research and Development Command; ATC = Air Training Command; R-W = Ramo-Wooldridge Corporation; SAC = Strategic Air Command; SAPO = Special Aircraft Projects Office; SecDef = Secretary of Defense; WDD = Western Development Division.

The Gillette Committee polled representatives from the relevant commands. It heard complaints from AMC, which did not support giving further authority to Schriever's group—not even its own Special Aircraft Projects Office. Earlier in the year, SAPO had proposed to form its own AMC counterpart to the Western Development Division, granting it substantial authority to perform logistics planning. The AMC Council rejected the request, stating that "if all major commands were to conform to this pattern, it would be tantamount to staffing a little Air Force all by itself out there on the West Coast." AMC instead created a Logistics Review Committee to act as an "information pipeline" from SAPO to HQ AMC. When the Gillette Committee asked for its input on how ICBM efforts should be organized, AMC's representative stated AMC's official view: The status quo was quite sufficient, except that the various organizations involved "should be instructed and impressed to restrain their activities to actions appropriate to the level." In addition, "all elements of all organizational levels involved . . . must be impressed with the same urgency to the program. . . ." Exhortation would sufficiently speed the process, according to AMC.[31]

The problem with the status quo, as the Robertson Committee discovered, was the long time delay inherent in established procedures. The committee found that the multiple approvals and reporting lines delayed the issuance of an industrial facility contract an average of 251 days from receipt of requirements. That simply was unacceptable. In consequence, AMC's preference for the status quo did not prevail.[32] Figure 9 shows the significant change in organization and reporting based on the Gillette Procedures.

The Gillette Procedures, approved by Secretary of Defense Charles Wilson on November 8, 1955, allowed WDD completely to bypass the Air Force for approvals and decisions and to funnel all ballistic missile decisions through a newly formed Ballistic Missile Committee in the OSD. Although it evaded ARDC and AMC for approvals and decisions, Schriever's organization needed to send information to those two Air Force commands. Schriever stated that "we had to give them information because they provide a lot of support, you see, so it wasn't the fact that we were trying to bypass them. We just didn't want to have a lot of peons at the various staff levels so they could get their fingers on it." The Ballistic Missile Committee would review an annual ICBM development plan, and the OSD would fund, present, and approve the ICBM program separately from the rest of the Air Force budget. The development plan included information on programming (linking plans to budgets), facilities, testing, personnel, aircraft allocation, financial plans, and current status. No longer could various Air Force and DOD organizations modify the program through piecemeal changes. Like the WDD, they had to work through the Development Plans and the Ballistic Missile Committee. The new procedures had the intended effect because by 1958 industrial facility lead time had been cut from 251 to 43 days at Air Materiel Command.[33]

The new procedures worked to Schriever's advantage and they relegated AMC, ARDC, and the operational commands to roles aiding the ICBM program but without authority to change it. The only good thing about the program from a parochial Air Force standpoint was that the completed missiles would eventually become part of Gen. Curtis LeMay's Strategic Air Command. Many in the Air Force did not take ballistic missiles seriously enough to make them worth fighting for. Col. Ray Soper, one of Schriever's trusted subordinates, noted that "the Ops [operational commands] attitude, at the Pentagon, was to let the 'longhairs' develop the system—they really didn't take a very serious view of the ballistic missile, for it was thought to be more a psychological weapon than anything else."[34] Those who did take missiles seriously supported the Gillette Procedures as the lesser of two evils:

[T]here was yet another argument in favor of the Gillette Procedures—kept within the Air Staff—and this was the fear that Secretary Gardner might succeed in making the Air Force Ballistic Missile Program another Manhattan Project, with himself as "czar." The Air Staff wanted ballistic missiles to evolve as a strategic weapon system and become completely integrated into the Air Force. The Special Procedures insured that this would

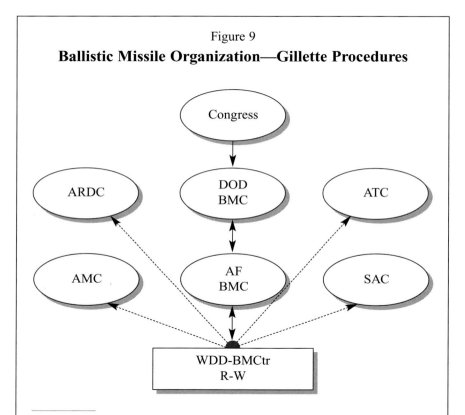

Figure 9

Ballistic Missile Organization—Gillette Procedures

Note: Solid lines show the direct chain of authority. Air Force commands have no authority over ballistic missile development, and the Air Staff has input only through the DOD Ballistic Missile Committee.

AF = Air Force; AMC = Air Materiel Command; ARDC = Air Research and Development Command; ATC = Air Training Command; BMC = Ballistic Missile Committee; BMCtr = Ballistic Missile Center; DOD = Department of Defense; R-W = Ramo-Wooldridge Corporation; SAC = Strategic Air Command; WDD = Western Development Division.

be so and served as a block to Gardner's ambitions, real or imagined.[35]

With the adoption of the Gillette Procedures, Schriever and Gardner garnered authority directly from the president, with a single approval of a single document to carry out the ICBM mission. Schriever's organization drew on the best Air Force personnel and services without that service interfering with his authority or decision processes, except through their representation on the Ballistic Missile Committee at the level of the secretary of defense. These new procedures represented the first full application of project management in the Air Force, an approach in which the project manager has both technical and budget authority for the project. Prior to this time, each project drew funds from several sources (budget line items) and so required separate justifications to the personnel in charge of each budget. The Gillette Procedures made the Air Force's financial and accounting system consistent with the authority of the project manager. With these procedures in hand, Convair and the contractors under control, and the Air Force's regular bureaucracy shunted out of the way, Schriever drove the ICBM program at full speed with little heed to cost.

Applying the System Concept

To proceed at full speed, Schriever's organization had to balance informality to gain speed with formal mechanisms to ensure political support, meet legal criteria, and achieve reliability for his missiles. While he worked to obtain sufficient authority and funding, Schriever's other essential task was to organize his triumvirate of organizations to oversee and direct ICBM development. To do so, he fell back upon concepts and processes developed since World War II, including the systems approach and its technical counterpart, systems engineering. The application of systems ideas to ballistic missiles within Schriever's command complex yielded a unique blend of methods tailored to rapid development of ICBMs.

First and foremost, the systems approach meant planning at the outset for the entire life cycle of the weapon. One of Schriever's first actions was to establish a centralized planning and control facility to support application of this idea. The WDD established its own local and long-distance telephone services, including encrypted links for classified information and teletype facilities. In the fall of 1954 Schriever and his staff developed a management control system under which the Air Force, R-W, and the associate contractors were required regularly to fill out standardized status report forms—shorter ones completed weekly and more thorough forms filed monthly. One of Schriever's officers had sole control of these reports and from them he updated the master schedules, which he placed on the walls of a guarded program control room. That room served a dual purpose as the place where managers quickly could assess the "official" status of the program and as a public relations tool that Schriever and his deputies used to show the program status and innovative management to visitors.[36]

The primary benefit of the management control system was the process of preparing the weekly and monthly status reports. The preparation required that managers collect and verify data, identify and report problems, and make recommendations for resolving them. Schriever instituted monthly conferences known as "Black Saturdays" for project officers to report difficulties. At these meetings Schriever and his top Ramo-Wooldridge and military staff reviewed the entire program and assigned all problems presented at the meeting to participants. Attendees endeavored to bring problems forward for solution instead of letting them be swept under the rug. As Schriever put it, "the successes and failures of all the departments get a good airing."[37]

Applying the system concept to ICBMs implied not only the parallel planning of all system elements but also their concurrent development. Schriever

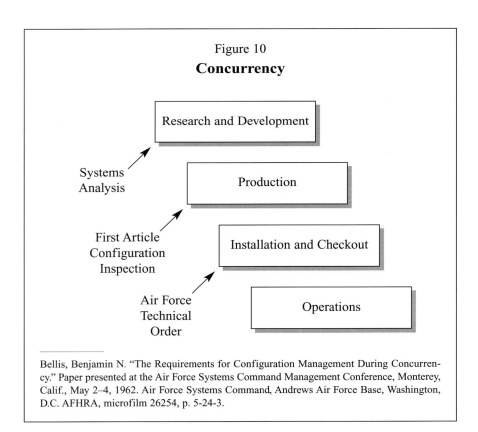

Figure 10
Concurrency

Research and Development

Systems
Analysis

Production

First Article
Configuration
Inspection

Installation and Checkout

Air Force
Technical
Order

Operations

Bellis, Benjamin N. "The Requirements for Configuration Management During Concurrency." Paper presented at the Air Force Systems Command Management Conference, Monterey, Calif., May 2–4, 1962. Air Force Systems Command, Andrews Air Force Base, Washington, D.C. AFHRA, microfilm 26254, p. 5-24-3.

called this "concurrency," a handy word that connoted his particularly rapid application of the systems approach (Figure 10). Concurrency simply meant that management telescoped several typically serial activities into parallel ones. In serial developments, research led in sequence to initial design, the creation of prototypes, testing, and manufacturing. Once manufactured, the operational units developed the maintenance and training required to use the new weapon. Under concurrency, these elements overlapped. Schriever did not invent the process but rather coined the term as a way of explaining it to outsiders.[38] Along with the program control room, concurrency gave his organization an aura of competence that helped protect it from detailed scrutiny.[39]

Schriever's version of concurrency combined concepts learned over the previous decade. Parallel developments such as the Titan had been practiced during World War II on the B–29 and Manhattan Projects. Centralizing management around the product instead of by discipline also had been used on these projects. The combination of ARDC's WDD and AMC's SAPO into a project-based office was a method applied since 1952. Schriever's use of R-W to perform systems analyses like that on Atlas' nose cone design also had been foreshadowed by RAND's development of systems analyses since the mid-1940s. What, if anything, was new?

The difference simply was that in the 1950s, with accelerated development undertaken in peacetime, Schriever had to explain his processes more than did his wartime predecessors. Schriever and others used the word to explain their strategy and management approach to Congress. As Secretary of the Air Force James H. Douglas later told Congress, "I am entirely ready to express the view that . . . you have to subordinate the expenditure . . . to the urgency of looking to the end result." Or as Trevor Gardner succinctly stated, "We have to buy time with money." The term "concurrency" helped explain and justify their actions to higher authorities.[40]

One of Schriever's first tasks was to evaluate and select contractors first for Atlas, and then for Titan and Thor. R-W performed the technical evaluations and gave their inputs to ad hoc teams of WDD and SAPO personnel. Those teams then jointly assessed management performance; financial, manufacturing, and development capabilities; past performance; security; and vulnerability. The AMC–ARDC committee selected those companies that should be asked to bid, evaluated the bids, and selected a second contractor for some subsystems. Selecting a concurrent contractor increased the chances of technical success, stimulated better contractor performance by threatening a competitive contract in case of poor performance, and kept contractors working while the Air Force made decisions. To speed development, the SAPO issued numerous letter contracts to get the contractors working, deferring contract negotiations until later. In January 1955 the SAPO formalized the ad hoc committees, which became the AMC–ARDC Source Selection Board. The board requested R-W's assistance as they deemed necessary.[41]

An Air Force SM–75 Thor intermediate-range ballistic missile is readied for launch at the Missile Test Annex, Cape Canaveral, Florida.

Schriever initially organized the WDD along functional lines with disciplinary divisions. Only in 1956 did the proliferation of projects lead him to create WSPOs for each project. These offices consisted of AMC and ARDC representatives, as required by the weapons system concept. Until that time, most work occurred through ad hoc teams led by officers who had the most experience in each area. For example, when WDD began to develop design criteria for facilities in March 1955, Schriever named his technical deputy, Col. Charles Terhune, team captain for the task. He also requested that R-W personnel assist in developing the criteria.[42]

Facilities planning, along with logistics, required a good grasp of how ICBMs would be operated. To begin understanding how this might be done, WDD contracted for studies through SAPO with RAND and with consulting companies Holmes and Arthur D. Little. In December 1954 WDD also initiated a series of meetings with AMC, Strategic Air Command, Air Force Headquarters, and other commands.

After the December meeting the AMC Council decided it needed quarterly reports from WDD to keep it abreast of events. Over the next six months bickering over reporting and support separated WDD from the Air Materiel Command planning groups, who needed the information for their personnel and logistics planning. AMC tried to maintain their usual procedures of performing the planning tasks at Wright Field, whereas WDD (and soon SAPO) accomplished their planning rapidly on-site, with little documentation or formality. AMC accused WDD of refusing to give the necessary data; the WDD accused AMC of a lack of interest. Disturbed because Schriever's crew had neither WSPOs nor Weapon System Phasing Groups (normally used to coordinate logistics), AMC had some reason to complain. As stated by Assistant for Development Programming Brig. Gen. Ben Funk, "the normal organizational mechanisms and procedures for collecting and disseminating weapon system planning during the weapon system development phase did not exist," leading to gaps in the flow of information necessary for coordination. By the summer of 1955 SAPO personnel at WDD made concerted efforts to pass information back to AMC and to bring AMC planning information into WDD.[43]

Schriever's need for speed led to extensive use of letter contracts through 1954 and 1955. Procurement officials in the SAPO and technical officers in the WDD realized that they needed to track expenditures relative to technical progress, but the rapid pace of the program and the lack of documentation quickly led to a financial and contractual morass. Complicated by the WDD's lack of personnel and the new process of working with R-W to issue technical directives, the contractual problems became a major headache for SAPO and AMC, and another source of friction between Schriever and AMC.[44]

The SAPO had authority to negotiate and administer contracts but initially lacked the personnel to administer them over the long term. Instead, SAPO personnel reassigned administration to the field offices of other commands "through special written agreements."[45] This complicated arrangement inevitably led to trouble. Part of the problem was the difficulty of integrating R-W into the management of the program. R-W had the authority to issue contractually binding "technical directives" to the contractors, but instead of doing so R-W personnel sometimes "used the technical directive as a last resort, preferring persuasion first through either periodic meetings with contractor personnel or person-to-person visits between R-W and contractor personnel." This meant that many design changes occurred with no legal or contractual documentation. Because officers in the SAPO did not have enough personnel to monitor all meetings between

Maj. Gen. Bernard A. Schriever (center) addresses employees gathered for the opening of the Western Development Division's new office complex near the Los Angeles International Airport in 1956. Standing to Schriever's left is Simon Ramo. The complex housed the ballistic missiles office of the Air Materiel Command and the guided missile research division of Ramo-Wooldridge Corporation.

R-W and the contractors and were not initially included in the "technical directive coordination cycle," matters soon got out of hand.[46]

The problem emerged during contract negotiations as SAPO procurement officers and the contractors unearthed numerous mismatches between the official record of technical directives and the actual tasks and designs of the contractors. As negotiations revealed the differences, actual costs spiraled upward, leaving huge cost overruns uncovered by any existing or planned funding. A committee appointed to investigate the problem concluded in June 1956 that "almost everyone concerned had been more interested in getting his work done fast than in observing regulations." It took the committee somewhat more than six months to establish revised procedures acceptable to all parties.[47]

The Procurement Staff Division of the Ballistic Missiles Office at Air Force Headquarters had to cope with the legal and financial mess. They insisted that "the technical directives [be] covered by cost estimates" because the annual funding from the Department of Defense was insufficient to cover the earlier-planned cost. Schriever's organization fought these regulations as "examples of the 'law's delay,'" but in the end had to give in. In November 1956 Schriever agreed to submit cost estimates, leading to new procedures in February 1957. To ensure that Ramo-Wooldridge and the other contractors documented technical directives, in October 1956 the guidance branch of WDD "began holding a contract administration meeting immediately after each technical directive meeting.... By January 1957 the Procurement Staff Division extended the practice to all contracts involving technical direction and to all technical direction meetings."[48]

Ramo-Wooldridge's role in systems engineering and technical direction for the ICBM programs was certainly one of the unique features of Schriever's application of the systems approach, and it was probably the one element that caused the most difficulties from an organizational and legal standpoint. Schriever believed in the necessity of close working relationships among R-W, WDD, and SAPO and he took a number of steps to enhance and protect this unusual organizational structure. First, Schriever and Simon Ramo agreed that R-W personnel

Maj. Gen. Bernard Schriever (left) confers with R-W managers Simon Ramo (center) and Louis Dunn in June 1956. Dunn left his position as director of the Jet Propulsion Laboratory to head R-W's Atlas effort.

should be placed in offices adjacent to those of their WDD counterparts. For example, Schriever's technical director Charles Terhune had his office next to the R-W technical director Louis Dunn. Schriever and Ramo were in frequent contact with each other regarding various issues.[49]

Because of the continuing sensitivity of Convair and other contractors to R-W's insider position, Schriever and the Air Force took a number of measures to deflect criticism and ensure effective working relationships. One critical action

Atlas missiles on the production line at the San Diego, California, plant of General Dynamics Corporation's Convair Division in 1958. Unlike aircraft, which required only a few test vehicles, missiles required an assembly line right from the outset.

was to remove R-W from legal and financial supervision of contracts. In November 1954 Schriever had told ARDC Commander Power that he would ensure all subcontracts R-W required for its studies and analyses would be issued and monitored by the SAPO to remove this potential target for criticism.[50]

The actual roles of R-W were not clear to Schriever's group as late as April 1955. Because of the continuing contractor problems, which included Convair and other contractors' refusals to put R-W on the distribution list for required documentation, Schriever directed Ramo to put together a briefing to describe for his officers and to the contractors the processes and tasks that R-W actually performed. In response, Ramo described R-W's organization and activities, thereby formulating one of the earliest descriptions of the practice of systems engineering and technical direction.[51]

Systems Engineering
from the Ridenour Report to Ramo-Wooldridge

Systems engineering had been mentioned in the 1949 Ridenour Report of the SAB that led to the founding of ARDC. Because of the Air Force's well-intentioned interest in component standardization across many aircraft and technologies to reduce costs, the components laboratories of AMC's Engineering Division had too much control over the development of the overall systems. The Ridenour group recommended that "the role of systems engineering should be substantially strengthened, and systems projects should be attacked on a 'task force' basis by teams of systems and component specialists organized on a semi-permanent basis." Who were these systems engineers? They were the aircraft project officers and other project engineers whose job was to integrate all components into a single aircraft, missile, or air defense system. The Ridenour Report recommended that these personnel be given substantially more authority and autonomy.[52]

The Air Force's response to the Ridenour Report with regard to systems engineering was to create the WSPOs and to give the ARDC and AMC project officers more authority. This met with resistance from the Wright Field engineers because it took away some of their authority. Modifying regulations to give project officers more authority helped somewhat, but overcoming existing biases required education and exhortation as well.

By early 1952 Doolittle, Putt, and Schriever were taking their show on the road to educate Air Force engineers and officers in the methods and benefits of the systems approach. Doolittle and Schriever went to Wright Field in January 1952 to explain the roles of ARDC and AMC and the systems approach espoused in the "Combat Ready Aircraft" study. Doolittle agreed that the split between ARDC and AMC had slowed aircraft development because the project officers were not always technically competent enough to answer industry's questions and so they had to go to the component engineers for answers. Doolittle told the Wright Field engineers that he was pleased to see the formation of the project offices, despite the delays, and told them that the solution to the problem was to train more qualified project officers.[53]

Longtime Wright Field engineer and manager Col. Marvin Demler was asked in late 1951 to study Wright Air Development Center's (WADC) management practices, and his report in February 1952 supported the new approaches. Demler believed that Wright Field's management had gotten too involved with the development of systems when their organization and expertise was in development of components. Because of the emphasis on component engineering, Demler thought that the project engineer had been overlooked and that the oversight led to the division managers' taking on too much system coordination work. Demler agreed with the Ridenour Report that the solution was a greater emphasis on systems planning by task teams who would have greater authority than they currently did.[54]

The assault on the old ways of doing business continued when Donald Putt was appointed commanding general of WADC. In April 1952 he spoke with the chiefs of the WADC laboratories about the changes involved in adopting the systems approach. He began by identifying industry resistance to strong project control, noting that the resistance was because "you in the laboratories prevented them from doing exactly what they wanted to do rather than what their customer, the Air Force, wanted done." Because the industry made money on the number of aircraft that rolled off the line, it didn't want WADC to interfere by requiring attention to quality. Thus identifying strong project control as one of the laboratories' strengths, Putt then admonished the WADC laboratory chiefs that "somebody has to be captain of the team, and decide what must be compromised and why. And that responsibility we have placed on the project offices." Under the old organization, the technical function "was anybody's baby," but under the new organization, WADC would have more technical clout through the project offices.[55]

Another aspect of the weapons system approach further reduced the authority of Wright Field's engineers. In his capacity as vice commander of ARDC, Putt issued a memo to the WADC commander in December 1952 describing the application of the system concept. The new procedures were to emphasize "integrated systems engineering," which would be "materially aided by giving a prime contractor systems design and engineering responsibility for the complete weapon." The Air Force was to retain veto power and monitoring capability because it "cannot escape its own responsibility for systems management simply by assigning larger blocks of design and engineering responsibility to industry." In addition, project offices needed to ensure that the contractor had systems engineering competence. Only if the prospective prime contractor did not have that competence should the associate contractor method be used. Despite Putt's admonitions about retaining final veto power, the use of prime contractors to integrate and deliver full systems was a further blow to the component engineers.[56]

The technical characteristics of missile systems placed even more responsibility with the contractors and the project offices. Because missiles were used only once and often required new, specially developed ground facilities, missile testing could not be performed exclusively by the Air Force, as was typical for aircraft. When the Air Force took delivery of aircraft prototypes for testing, WADC engineers and Air Force test pilots from Air Proving Ground Command performed a series of tests on the prototypes. Without pilots and with only one flight test opportunity per missile, testing required that a series of missiles be fired. Engineers had to incorporate lessons learned from each test into the production line and into missiles in the manufacturing process. This in turn implied a much greater involvement of the development contractor in the testing process, and a relative diminution of the responsibilities of Air Force officers. To verify missile performance, officers monitored tests conducted by the contractors instead of performing the tests themselves.[57]

On Schriever's ICBM programs, WADC officers were shut out of the process even further because Ramo-Wooldridge gave WDD the technical support that WADC component engineers normally gave to the project officers. With his emphasis on speed over regulations, Schriever consistently pressed to acquire the most talented people and to colocate them in El Segundo, California (they all moved from Inglewood in 1955). He was not willing to use WADC's expertise in Dayton any more than that of AMC unless he could have their people located at his facility. Although Schriever rejected the prime contractor approach because of Convair's apparent lack of systems engineering capability, he believed that R-W's civilian scientists and engineers had systems engineering skills that the Air Force lacked.

Ramo-Wooldridge formed its Guided Missile Research Division (GMRD) in 1954 to handle the technical aspects of the ICBM programs. With Simon Ramo heading the division and Louis Dunn his technical deputy in April 1955, the GMRD had five departments—aeronautics R&D, electronics R&D, systems engineering, flight test, and project control. While the aeronautics and electronics departments concentrated on subsystems and components, respectively, the systems engineering, flight test, and project control departments performed the bulk of ICBM integration tasks.[58]

Technical direction of the contractors took place through monthly formal meetings and numerous informal meetings as appropriate to the situation. In the formal monthly meetings, R-W project control personnel chaired the meeting, set the agenda, recorded minutes, and presented the current schedules and decisions. Based on the results of these meetings, the project control department issued technical directives, work statements, and contract changes. Cognizant officers in WDD reviewed the technical directives and changes to work statements. Work statements and contract changes were submitted to the SAPO and officers there then issued contractual changes and approved work statement modifications. Informal meetings were for "information only" although such information would be coordinated between WDD and R-W as necessary. As discussed previously, it took more than a year before coordination between R-W and the SAPO effectively matched technical directives to work statements and contract changes. The project control department handled official plans, schedule work statements, cost estimates, and contract changes.[59]

The systems engineering department concentrated its efforts on major design interactions, the electrical and structural compatibility between subsystems and contractors, and issuance of top-level requirements. One good example was the nose cone trade study that led to the reduction of Atlas's weight by half. Another example was an assessment of Martin Company's trajectory analysis. They found that Martin's trajectory was less than optimal and by modifying it R-W engineers increased the Titan's operational range by six hundred miles, an amount equivalent to saving 10 percent of its weight. Numerous other examples of R-W technical personnel catching problems made by the contractors were documented as both Schriever and R-W needed proof that R-W was performing a valuable service for the Air Force. R-W's systems engineers also performed laboratory experimental work when they needed more information, analyzed intelligence data on Soviet tests, and programmed the early missiles. As noted by one critic of R-W, they often focused on double-checking the contractors and avoiding "errors, mistakes, and failures."[60]

By October 1956 WDD and R-W had come to a legal agreement of what systems engineering entailed, sufficient to define in some detail the actual tasks involved. The WDD–R-W agreement defined systems engineering in terms of three functions:

1. The solution of interface problems among all weapon system subsystems to ensure technical and schedule compatibility of the systems as a whole.
2. The surveillance over detailed subsystem and overall weapon design to meet Air Force required objectives.
3. The establishment and revision of program milestones and schedules, and monitoring of contractor progress in maintaining schedules, consistent with sound technical judgment and rapid advancement of the state of the art.

Based on those functions, the statement of work for R-W included the following items:

- conducting research studies required to carry out technical evaluations and systems analysis
- conducting experimental investigations with approval of the Air Force
- planning and conducting systems engineering
- preparing and maintaining portions of the R&D plan as assigned by the Air Force

- preparing, selecting, analyzing, and recommending specifications
- conducting essential studies and making recommendations concerning subsystem and component research, test, training, and operational programs, anti-ICBM countermeasures, and standards
- preparing work statements
- performing technical evaluations
- performing technical direction
- performing studies to help weapon deployment and operations
- facilities planning
- performing studies and preliminary systems engineering for the Advanced Reconnaissance System [a proposed spy satellite]
- maintaining ability to monitor science to aid ICBM development
- maintaining the control room.[61]

From 1953 through 1957, Ramo-Wooldridge's role grew dramatically. Starting out by documenting the proceedings of the Teapot Committee's deliberations, R-W then acquired a contract with Schriever's new organization to perform long-range studies of ICBMs, to assess potentially helpful new technologies, and to help the WDD set up and operate its new facilities in Inglewood and then El Segundo. Its funding grew from $25,494 through June 1954, to $833,608 from July 1954 through June 1955, and to $10,095,545 from July 1955 through June 1956. R-W managers allocated their growing budgets the proportions shown in Table 2. As R-W's competence grew and the results of its studies and technology assessments proved beneficial, at the Teapot Committee's recommendation Schriever expanded its role to include systems engineering and technical direction. In essence, R-W looked over the shoulders of all of the contractors and double-checked what they did; controlled the detailed specifications, schedules, and other paperwork; and kept a look out on the technical horizon for new technological solutions. As Schriever himself later admitted, R-W did for the WDD what Wright Field and its component engineering expertise did for aircraft development. For the first couple years of expansion, R-W's services were indispensable to the WDD, cutting program costs and improving the performance of the WDD's stable of ballistic missiles. The real proof, however, would come when flight tests began—and for the WDD and R-W that would prove to be a rocky experience.[62]

Testing Concurrency

Although Schriever could and did shrug off cost factors and bureaucratic snafus, he could not ignore technical problems. Unfortunately for him and his desire to reduce bureaucracy, concurrency complicated technical problems instead of simplifying them. He and his team discovered these problems when the contractors and the Air Force began ballistic missile testing in late 1956 and 1957.

Table 2

Ramo-Wooldridge Budget Expenditures

Estimated May 1957

Item	Percentage of Budget (%)
Systems engineering and technology direction	35
Supporting services for WDD	24
In-house experimental studies	10
Auditing of contractor's engineering	9
Aid to contractors on specific tasks	9
Operational studies for WDD	7
Management support to WDD	6

WDD = Western Development Division.

Source: Ramo-Wooldridge interoffice correspondence, Simon Ramo to Dean Wooldridge, subject: R-W Contract as a Function of Total USAF Ballistic Missile Programs, May 13, 1957, item 9, AFHRA, microfilm 35267.

The Air Force and industry based their processes on those developed in the aircraft industry since the 1920s. One of the great advantages of sequential development as practiced before World War II was that problems could be found during a long process of prototype inspections and flight tests. Armaments and electronic equipment could be added to the aircraft later because of their loose connection to the airframe and its performance. Wright Field engineers typically would add electronics and armament after aircraft manufacturers delivered a prototype and then could redesign the prototype based on actual hardware to produce the preproduction model, which Air Corps personnel tested before ordering a large number of airplanes. Detailed planning was less important when you could make changes to accommodate new equipment as necessary.

In the rush to bring missiles into the operational inventory, Schriever ordered all elements designed at the same time. This concurrent development had its complications because if engineers found a design problem in a component, that problem could affect not only other elements of the flight vehicle but potentially production tooling, other flight vehicles already built, ground equipment, and operations training. Unfortunately, because ballistic missiles were still in their infancy design changes were a way of life that caused frequent ripple effects in numerous other organizations. Not surprisingly, that caused tremendous cost overruns as production tooling was thrown out, missile bases were built and rebuilt, and retraining followed training. Under concurrency, problems dealt with in a leisurely way in serial development had to be found and fixed rapidly or their ramifications could be technically and financially devastating.

In a field that changed as rapidly as aircraft R&D and production, the Air Force had developed a number of methods to detect problems before ordering aircraft into production. Over the years, Wright Field engineers and procurement officers had created a system of reviews, approvals, and procedures for aircraft contracting, development, and testing. At every phase of aircraft development, Wright Field officers and civilians reviewed and approved specifications, designs, hardware, and tests. They governed contractual requirements with general specifications levied on all equipment and detail specifications levied on unique components. These inspections, reviews, and approvals occurred not only at the break points between studies, prototype X-model, preproduction Y-model, and full production, but also while the aircraft contractor developed specifications, analyses, designs, hardware, and supporting facilities. Once an aircraft was built, Wright Field officers conducted a series of tests to check the aircraft's performance, contractual compliance, all-weather capability, structural integrity, and operational capability. These numerous approvals, inspections, and tests ensured that the contractor supplied a safe aircraft with sufficient performance, albeit at a higher cost and on a slower schedule than without these constraints. Although Schriever wanted to develop ICBMs as rapidly as possible, he, his Air Force team, and the industrial designers did not want to throw out the means to detect design problems.[63]

Pilotless, one-time-only missile testing differed a great deal from aircraft testing. As noted previously, the missile manufacturer was involved with testing to a far greater degree than was the aircraft manufacturer. In addition, the Air Force's aircraft testing procedures and facilities did not apply very well to missiles. For aircraft, the Air Force had developed its "unsatisfactory report system" whereby test pilots, crew members, and maintenance personnel reported problems, which were then relayed to Wright Field engineers for analysis and resolution. With missiles, there were no pilots, crew members, or maintenance personnel during missile development testing.[64]

Because each missile was destroyed on completion of its test flight, flight tests needed to be minimized and preflight ground testing maximized, with Air Force officers witnessing contractor testing. The high cost of flight testing made simulation a cost-effective option, along with the use of "captive tests" in which the rocket would be tied down to the launch pad while it was fired. R-W engineers calculated that for ICBMs to achieve a 50 percent probability of successful operations in wartime conditions, they should achieve 90 percent flight success in ideal testing conditions. In turn, this required that each subsystem of the missile could fail only 2 to 3 percent of the time. With such a limited number of flight tests, this failure rate could not be proven statistically. Instead, R-W developed a philosophy of thorough checkout and testing for all components and subsystems prior to missile assembly. Flight tests would be reserved for observing interactions among subsystems and for studying overall missile performance. Initial flight tests would start with the minimum configuration, that is, with only the airframe, propulsion, and autopilot. Upon successful completion of those tests, engineers would add more subsystems for each subsequent test until the entire missile with warhead was tested successfully. Only then would the development phase of the program be completed.[65]

By 1955 each of the military services recognized that rocket reliability was a difficult problem, and ARDC sponsored a special symposium to discuss the subject.[66] Statistics from testing organizations found that two-thirds of the failures were caused by electronic components, such as vacuum tubes, wires, and relays. Electromagnetic interference and radio signals caused a significant number of failures, and about 20 percent of the problems were mechanical, dominated by hydraulic leaks.[67]

Atlas's test program proved to be no different. The first two Atlas tests in mid-1957 ended with engine failures but the third succeeded, leading to a record of three successes and five failures for the initial Atlas A test series. Similar statistics marked the Atlas B and C series tests between July 1958 and August 1959. For Atlas D, the first missiles in the operational configuration, reliability improved to 68 percent. Of the thirteen failures in the Atlas D series, four were caused by personnel errors, five were random part failures, two resulted from engine problems, and two occurred because of design flaws uncovered through testing.[68]

Engineering experimentation was more critical to fixing these problems than were scientific analyses, a fact propounded by the WDD's senior rocket engineer, Col. Edward Hall. One of the few Air Force officers experienced in rocket propulsion, Hall accepted assignment to Schriever's organization only under the condition that R-W would have no technical direction authority over him or his contractors. Schriever agreed because he needed Hall's expertise but, according to Hall, Schriever never realized that "the great scientists" of R-W "couldn't do the job" because the ballistic missile was not "a scientific job."[69]

Solving missile reliability problems proved to be extraordinarily difficult. In 1960 two accidents dramatized the reliability problem. In March an Atlas missile exploded, destroying its test facilities at Vandenberg Air Force Base on the California coast. Then in December the first Titan I test vehicle blew up along with its test facilities at Vandenberg. Both explosions occurred during liquid propellant loading, a fact that helped spur the development of the solid-propellant-based Minuteman missile. With missile reliability continuing to hover at the 50 percent range for Atlas and at a somewhat better 66 percent for Titan, concerns increased both inside and outside the Air Force.[70]

**A Titan I interconti-
nental ballistic mis-
sile, raised from its
underground silo,
stands ready for
inspection at the
Operational System
Test Facility at Van-
denberg Air Force
Base. The Titan was
a fully redundant
system to replace
Atlas in case that
program did not
succeed.**

Although the Air Force told Congress that they estimated missile reliability
at near 80 percent, knowledgeable insiders thought otherwise. When he looked
into the figures given to Congress, Col. Ray Soper, one of Schriever's deputies,
called the 80 percent figure "optimistically inaccurate" and estimated the true reliability at 56 percent in April 1960.[71] That same month Brig. Gen. Charles Terhune, who had been Schriever's technical director through the 1950s, entertained
serious doubts. While trying to find ways to improve ICBM reliability, Terhune
believed that the most difficult problem might be that "it is very hard to outline a
test program for this type of activity which causes the participants to feel they
are truly accomplishing something."

> The fact remains that the equipment has not been exercised,
> that the reliability is not as high as it should be, and that in all
> good conscience I doubt seriously if we can sit still and let this
> equipment represent a true deterrence for the American public
> when we know that it has shortcomings. In the aircraft program
> these shortcomings are gradually recognized through many

flights and much training and are eliminated if for no other reason, by the motivation of the crews to keep alive but no such
reason or motivation exists in the missile area. In fact, there is
even a tendency to leave it alone so it won't blow up.[72]

SAC weighed in with its own complaints. Ever skeptical about missiles because they might make SAC's strategic bombers obsolete, SAC officers informed
Air Force Headquarters in October 1960 that since acquiring the first Atlas missiles one year before, they have had "zero probability" of launching an Atlas and
having it hit a target. SAC recommended a dramatic increase in the size of the
B–52 fleet.[73]

ICBM reliability problems drew attention in the form of Air Force and congressional studies and investigations. An Air Force board with representatives
from ARDC, AMC, and SAC reported on the problems in November 1960, blam-
ing inadequate testing and training and insufficient configuration and quality
control. They recommended additional testing and process improvements, which
they called "Golden Ram." In December 1960, after the dramatic Titan explosion, Secretary of Defense Thomas S. Gates, Jr., requested a DOD-level investigation, ultimately performed by the Weapon Systems Evaluation Group, a systems analysis organization within the OSD. Another study by the director of
defense research and engineering, completed in May 1961, further criticized the
rushed testing schedules. Finally, in the spring of 1961 the Senate Preparedness
Investigating Subcommittee held hearings on the issue. They concluded that testing schedules were too optimistic. With technical troubles continuing, its own
officers ranging from concerned to hostile, and Congress putting pressure on
them, the West Coast complex had to find ways to make ICBMs operationally
reliable. To do so the Air Force and R-W created new organizational processes to
find problems and ensure high quality. Their success would vindicate their organizations and made their methods the cutting edge of Air Force management.[74]

Responding to Failure: The Creation of Configuration Control

As cost overruns and technical failures became obvious, the informal methods used by WDD and R-W became liabilities. Solutions to ICBM engineering
problems required rigorous processes of testing, inspection, and quality control.
In turn, these required better organization, tighter management, and improved
engineering control.

Reliability problems were the most immediate concern and the WDD and
AMC began their improvement efforts by collecting failure statistics. AMC had
required Atlas primary contractor General Dynamics to begin collecting logistics
data in late 1955, including component failure statistics. In 1957 AMC extended
this practice to other contractors, and later placed this data in a new, centralized

electrical data processing center.[75] Using methods of mathematical statistics, Ramo-Wooldridge scientists and engineers rationed a certain amount of "unreliability" to each element of the vehicle, backing up the allocations with empirical data. Using Air Force requirements as the overall reliability goal, they apportioned proportionately the required reliability levels as component specifications.[76]

Atlas program officers also developed an extensive empirical program to evaluate and test each vehicle. Starting on the operational vehicles of Atlas D, scientists and engineers at Space Technology Laboratories (STL, the successor to the GMRD of Ramo-Wooldridge) began a program to search for critical weaknesses. The program emphasized environmental and functional tests during which engineers stressed components "until failure was attained." They ran a series of "captive tests," holding down the missile while firing the engines. All components underwent tests that checked environmental tolerances (temperature, humidity, and so forth), vibration tolerance, component functions, and interactions among assembled components. These tests required the development of new test equipment, such as vacuum chambers and vibration tables, where all of the missile's components could be tested to withstand the flight environment well before they ever flew. By 1959 the Atlas program also included tests to verify operational procedures and training. STL personnel created a failure reporting system to classify failures and fed the data into the central database to perform statistical analyses.[77]

Buildings used by the Space Technology Laboratories, the successor to R-W's Guided Missile Research Division, in July 1959.

Engineers found that a number of test failures resulted from mismatches between the design of the missile and the actual hardware configuration of the missile on the launch pad. In the rush to fix problems, the launch organization, contractors, or Air Force had made modifications to missiles without documenting that they had done so. To fix that problem, STL personnel and Air Force officers developed configuration control, a reporting procedure that tracked and connected changes in missile design to the changes in the missile hardware. Because hardware changes often involved manufacturing and launch processes, configuration control soon came to control changes of these processes as well.[78]

Although inspired by problems endemic to the ballistic missile programs, configuration control drew from the Boeing Company's commercial and military aircraft programs. Boeing's experience came to the BMD–STL complex through Boeing's participation in the Minuteman program as the contractor responsible for assembly and testing. Boeing's quality assurance procedures used the following five control tools:

- formal systems for recording technical requirements
- a product numbering and nomenclature system for each deliverable contract item
- a system of control documents with space for added data on quantities, schedules, procedures, and the like
- a change processing system
- an integrated records system.

In addition, Boeing used a "change board" that ensured that all affected departments reviewed any engineering or manufacturing change and committed appropriate resources to effect it. Although Boeing did not highlight this in their 1958 proposal, the Air Force soon saw the importance of this process innovation and elevated it to "a separate management discipline, organized and staffed accordingly." Boeing's process included identification, change, and accountability control over all designs and corresponding hardware.[79]

Configuration control was a successor to the concept of the "design freeze," followed by engineering change orders. The design freeze was an important engineering milestone in aircraft development, the point in time when engineers stopped making design changes so that an actual piece of hardware could be built to match that design. Once they froze a design, engineers or operators could make changes only by submitting a formal engineering change request. This was necessary because any further design changes would then also involve changes to hardware already built and possibly to production facilities as well. On the Thor program R-W earned praise from Air Force officers for their thorough preparation that allowed them to freeze the design quickly.[80]

Configuration control provided BMD officers and STL engineers with the means to coordinate and control changes, ensure the compatibility of designs and hardware, and control and predict costs and schedules. The key to configuration control was creating a formal change board (called the configuration control board) with representatives from all organizations and a formal system of paperwork that linked specifications, designs, hardware, and processes. Although they started by linking design drawings to hardware, BMD officers and STL engineers soon realized that by expanding configuration control to include specifications and procedures, they could control the entire development process.

Through the mechanism of configuration control, specifications could be linked to designs, designs to hardware, and hardware to testing and operational procedures. Through the configuration control board, the engineering design groups and committees proposed their changes to the formal program hierarchy as represented on the board. Air Force officers soon linked the configuration control board results to contracts. By the early 1960s the military required all changes to include estimated increases to cost and delays to the schedule. When the change board approved a change, the government contractually committed itself to funding the changes, and to defending the new costs and schedules. Conversely, the contractor would be liable for the costs if the change was not approved by the board and formalized by a contract change. Configuration control became the most important tool linking engineering communication and design, management hierarchy and authority, and contractual links between the government and its contractors.[81]

The Air Force established configuration control in the fall of 1959 on Minuteman, and soon thereafter made it part of the other projects in the BMD–STL complex. The Atlas configuration control board began with Atlas D on missile 33D, which flew on September 29, 1960.[82] The complex created its first general regulations and guidelines for configuration control in 1962.* Although it was expensive, the Air Force vigorously promoted configuration control because of its utility in linking engineering, management, and contractual matters, and its capability to track costs and schedules.

Along with configuration control, STL also adopted more formal methods to handle multiple projects. By the early 1960s STL and the Aerospace Corporation[83] had evolved their coordinating role into a set of procedures called System Requirements Analysis (SRA). SRA was a procedure for controlling technology development through the control of requirements. Requirements were (and are) specific statements of program objectives. For example, at the highest level a requirement would be written to develop a ballistic missile system to deliver a one-megaton payload at a distance of five thousand miles with an accuracy of one mile. This statement would be broken into at least three statements at the next level. These three in turn would then be broken down into numerous requirements to create hardware components, operating procedures, and so on. Major

---

*The Aerospace Corporation was a nonprofit corporation set up to do the same functions as STL had done. The formation of Aerospace will be discussed later in the text.

programs derived thousands of implicit or explicit requirements corresponding to thousands of components and procedures. SRA made the design traceable to requirements by generating requirements at greater levels of detail through the development of the system.[84] Ramo-Wooldridge also developed common interface specifications and "change control procedures" to guide contractors. All of these documents supported the expansion of the configuration control system.[85]

Schriever's team of the BMD, the Ballistic Missile Office (the successor to the SAPO), and STL applied formal processes as they developed them to the Atlas, Titan, and Thor programs. However, by 1957 a solid-propellant alternative to the troublesome liquid-fueled ballistic missiles had become feasible. Colonel Hall, Schriever's controversial propulsion expert, had been studying solid-propellant technology for some time, and in late 1957 he barged into Technical Director Charles Terhune's office demanding that Terhune take some time to listen to what he had to say. After hearing him out, Terhune got hold of Schriever and had Hall present his ideas again. Given the importance of what Hall was saying, Schriever and Hall then went to Air Force Headquarters and won approval for what became known as the Minuteman program.[86]

What made Hall's presentation so compelling were the significant advantages of solid-propellant rockets and his extensive body of evidence to show that they could be made to work on a large scale. He pointed out that solid-fuel-propelled rockets did not have the costly, time-consuming, and dangerous liquid-

An Atlas D missile lifts off from Cape Canaveral, Florida, in an October 1960 test flight. The Atlas' early reliability hovered in the 50 percent range in the late 1950s, but they improved significantly thereafter with better management and systems engineering methods.

Shown inspecting the communications panel in an Air Force Minuteman launch control car are, from left to right, Col. Samuel Phillips, director of the Minuteman program office; Maj. Gen. O. J. Ritland, commander of ARDC's Ballistic Missile Division; and Lt. Gen. Bernard A. Schriever, ARDC commander.

propellant loading procedures needed for the current stable of liquid-propelled ICBMs. Although liquid-propelled rockets had higher performance, it took from several hours to several days to prepare them for launch. Solids, on the other hand, could be launched within seconds because, once loaded with propellant and placed in their launch configuration, the missiles were ready to go with the push of a button. For a nuclear deterrent that had obvious advantages. Hall started the program but he was too abrasive and unpredictable to manage it over the long run. Instead, Terhune recruited Col. Samuel C. Phillips to manage Minuteman.†

---

†The evidence points to origination of the idea of tying financial controls to change control on the Minuteman project, but does not definitively pinpoint it to Boeing or the Air Force. What is clear is that Phillips quickly saw the importance of the idea and promoted it vigorously.

One of the pads at the Air Force Missile Test Center from which three-stage, solid-fuel-propelled Minuteman ICBMs were launched in the missile's development program.

Phillips turned out to be an excellent manager, and he was equipped with all of the tools developed by Schriever's team, Ramo-Wooldridge, and system integrator Boeing. Boeing brought with them their ideas for configuration control, which Phillips quickly expanded to become an all-purpose managerial tool. Engineers needed change control to be able to communicate engineering changes to their counterparts on other subsystems, and work out any ripple effects. For example, if a power engineer had to change his subsystem's voltage from 5 volts to 7 volts, then all components and subsystems that use power had to change as well. Typically the project's chief engineer coordinated changes, ensured communication with all relevant subsystem engineers, and gave final approval or rejection of changes.

Sometime between 1958 and 1960, managers—probably those at Boeing or possibly Minuteman project manager Samuel Phillips—realized that configuration control was precisely the tool necessary for managers to gain financial as well as technical control over the project.[87] The idea was simple. All a project manager had to do was compel engineers to give cost and schedule estimates along with the technical change. If the engineer did not give the information, the change would be rejected. By requiring this information, the project manager could predict the cost profile of the project along with the schedule. This also allowed the manager to track the individual performance of each engineer or group of engineers because they could now be held accountable for the estimates that they had given. Managers made the process concrete by tying it to specific configurations of the hardware design, and eventually to the hardware itself. In addition, it was an excellent contractual tool because the procurement officials and industry managers could write contracts against specific design configurations and could negotiate cost changes based on each approved change. Phillips and others transformed configuration control into "configuration management," a critical managerial tool to control the overall R&D process.

Through configuration management, and because solid-fuel-propelled rockets avoided the dangers and reliability problems of liquid-propellant leaks, turbines, and other hydraulic and mechanical components, Minuteman boasted an enviable development record, coming in at cost and on schedule. The Air Force's improved procedures also made its liquid rockets more reliable. Because of Minuteman's much greater reliability and its launch-on-demand capability, the Air Force phased out liquid-propellant missiles as weapons through the 1960s. Their higher performance, however, made them excellent satellite launchers, in which role they and their descendents performed admirably. In their capacity as launchers, the Atlas, Titan, and Thor-Delta vehicles all attained greater than 90 percent reliability from the mid-1960s through the 1990s.

In this February 1961 photograph, a Minuteman I intercontinental ballistic missile leaves the launch pad at Cape Canaveral, Florida. Minuteman director Col. Samuel Phillips brought this project to completion on time and under budget using the method of configuration control.

The USAF and the Culture of Innovation

Configuration management, along with further attention to quality through thorough inspections, training, and associated documentation, has become an organizational pillar of the Air Force's management system. Its importance can hardly be overstated. Managers from the turn of the twentieth century through the 1950s had been searching for a way to predict R&D costs and control scientists and engineers. Configuration management became the primary means of this control on development projects. One reason that it became so important was that accountants and lawyers could tie technical modifications to contract modifications, including costs. This became a critical contractual mechanism to industry in the cost-plus environment of high-tech R&D. To understand the relationship with industry more fully, we must now consider how industry responded to Air Force application of the systems approach to aircraft and missiles.

The Systems Approach in Industry

While Air Force officers fought to advance government research and development, industry developed the actual hardware. Because the aircraft industry depended heavily on the military for the development of new products and for production contracts, changes in the Air Force's organization and procedures had significant effects on industry as well. In trying to make a profit in the boom-and-bust defense industry, industrial managers grappled with government directives and technical problems and eventually they arrived at a variation of the military's model for technical organization: matrix management.

Industry's primary concern was to acquire production contracts, which is where they made their profits. Research and development were necessary means to compete for new production contracts. Because the Army and Navy provided armament and electronics as needed, the aircraft companies had to build not complete systems but only the airframe and flying controls. Industry and the Air Corps interacted in a "sequential" manner whereby industry delivered products for testing by the government.[88]

During World War II, aircraft companies expanded dramatically to produce the thousands of aircraft required by the Army and Navy. Even while expanding, these companies tried to plan for the postwar period in which a large surplus of wartime aircraft would be available for purchase by commercial companies and new aircraft contracts would be scarce without wartime demand. Although jet engines and rockets appeared promising, contracts would be fewer and smaller than during the war.

After the war the aircraft companies shrank accordingly and observed the services' organizational and technological changes with interest. In the new missile programs and for the more complex aircraft such as the B–29, aircraft companies built entire systems that included ground equipment, armament, and electronics. These experiences led the aircraft companies to reorganize their own efforts around the complex new products. When the Air Force reorganized on a systems basis with greater emphasis on R&D in the early 1950s, aircraft contractors were well prepared to adjust, if they had not done so already. Each company, with a number of complex projects underway at the same time, had to reconcile the new project-based organization with their traditional discipline-based, functional structure.

The Martin Company developed one of the first project management organizations in the years 1952–53. William Bergen, an engineer who had worked in the late 1940s and early 1950s as head of the "pilotless aircraft group" and the Naval Research Laboratory's Viking rocket, was an early promoter of "systems management." As he described it in a 1954 *Aviation Age* article, "Within the company we have created a number of miniature companies, each concerned with but a single project. The project manager exercises overall product control—in terms of an organization of all skills." Martin quickly implemented Bergen's innovation and expanded it to "cover all functions from design through manufacturing and distribution." The Martin Company's systems approach included three elements: systems analysis to determine what to build, systems engineering to design it, and systems management to build it. When called on by the Air Force to build the Titan, they were already well on their way to implementing the systems approach.[89]

Another example was McDonnell Aircraft Company's F–4 Phantom program. In the 1940s McDonnell designed aircraft by committees staffed with engineers from its functional departments, with owner and president J. S. McDonnell arbitrating disputes. When the Navy awarded the company a sole-source contract to develop the F–4 in 1953, the company gave responsibility for it to bright young engineer David Lewis to "create an air of certainty around their inchoate proposal." Lewis was the first McDonnell employee given the title of project manager. Most of the engineers reported to functional engineering departments such as aerodynamics, structures, hydraulics, and electronics. Lewis assigned three project engineers outside of the functional departments' jurisdiction to make design decisions while he built the project organization and acquired resources. Within the Navy, a desk officer ran the F–4 program.[90] He and the four McDonnell project engineers formed a simple project management organization for the F–4 program.

Project management involved separating engineers from their functional departments and having them report directly to a project manager whose sole task was to run the program. The government contracting organization set up a similar structure, initially with a single person to manage the program. As projects grew the number of managers and engineers also grew, but all of them reported to project managers instead of to functional department managers. As stated by a business school professor in 1962, *"the primary reason for project management organization is to achieve some measure of managerial unity,* in the same way that physical unity is achieved with the project" (Davis's italics).[91] Project management organized around the structure of the product, not the structure of the parent institution.

With numerous systems under development in the mid-1950s, military contractors faced the problem of developing several of them concurrently. Confronted with a few massive, rapidly paced military programs, the old line-and-staff organization no longer was adequate. Communication lines across functional departments became too long for effective coordination. As stated by H. F. Lanier, a project engineer for Goodyear Aircraft's aerophysics department,

> The problem can perhaps be best illustrated by considering the difficulties of trying to fit a number of creative people into the precise and orderly line organization shown in [Figure 11]. Under this plan, all work is thoroughly organized and all assignments rigidly controlled. Each individual has a definite area to cover, definite data to work with, and a schedule to meet. He also has a boss who tells him what to do and subordinates whom he tells what to do. This organization once set up is soon limited to the creative output of a few men who lead. Any innovation is difficult to introduce because it requires detailed instruction at all levels.[92]

Lanier concluded, "The major step is somehow to break down the long lines of communication."[93] In the short term, organizations used ad hoc means. These were insufficient over the long term and for large projects:

> The usual solution was to allow a great deal of "co-ordination" and "liaison" to be handled informally. Effectively, supervisors unleashed their men and gave the program general direction but let detailed instructions be formulated after the fact. The loose method has been reasonably successful. The next obvious step is to attempt to systematize the process.[94]

Often the first attempt at systematizing was to form committees of the functional supervisors, but this did not work once meetings became too large or too frequent. "Usually the committee members are also line supervisors and hence can meet only for a fraction of the time required for efficient system development. In other words, actual development by a committee is employed most effectively on an occasional relatively huge problem. When large systems problems are the prime business, then a permanent fix must be made."[95]

For this purpose, Lanier stated that "the solution seems to be a committee of project or systems engineers—individuals trained to be jacks of all trades, and who are relieved of line responsibility for administering operating sections. . . ." He believed that the "project engineer is a feature as old as engineering. Groups

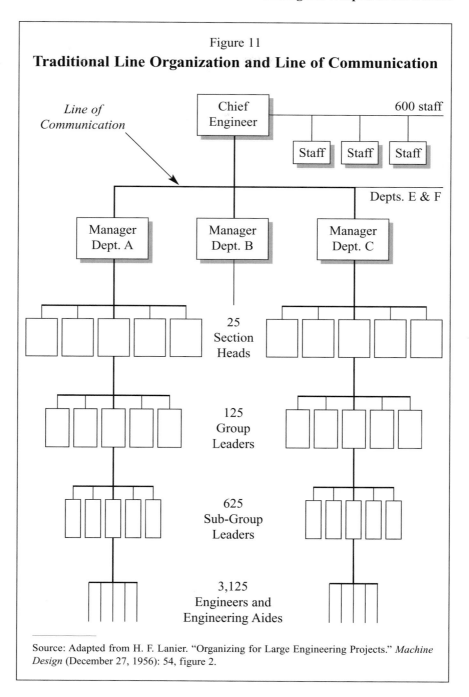

Figure 11

Traditional Line Organization and Line of Communication

Line of Communication

Chief Engineer

600 staff

Staff Staff Staff

Depts. E & F

Manager Dept. A

Manager Dept. B

Manager Dept. C

25 Section Heads

125 Group Leaders

625 Sub-Group Leaders

3,125 Engineers and Engineering Aides

Source: Adapted from H. F. Lanier. "Organizing for Large Engineering Projects." *Machine Design* (December 27, 1956): 54, figure 2.

of project engineers working in team effort under a project management is a little new." Project management aligned engineers to the project but left undetermined their relationship to the rest of the organization.

Lanier recognized that engineers had relationships both to the project and to the rest of the organization where the engineer must go when the project ended. Engineers therefore reported both to project and line management. Lanier called this dual reporting the "project-line combination organization." This form, he said, "existed in various forms for some time, usually as a special purpose, temporary thing. Now enough work of the large systems nature is under way to warrant the formation of permanent establishments geared to development of large systems." The new organizational form had a "two-dimensional" or "matrix" structure. Lanier noted that "several companies are experimenting with the arrangement [see Figure 12]. Here specialized creative engineering groups are given a two-dimensional supervision." In the new organization, the line manager and task manager both had roles in managing the "working group."[96]

The evolution of General Dynamics' Astronautics Division, responsible for the Atlas missile, typifies the organizational changes brought on by the company's involvement in several complex military projects. For most of the 1950s, Atlas was run as a single-project organization. Through its early years after the war, Atlas (then known as Project MX–774) was directed "by a project engineer who was assigned a small team of designers and technical specialists plus an experimental shop for fabrication of the hardware."[97] By 1954 one year after the acceleration of Atlas, General Dynamics' Convair Division reorganized the project around the program office and had a force of 300 personnel, mostly engineers. In 1955 the company created the Astronautics Division to carry out the work of the Atlas program. By 1958 the workforce had increased to 9,000, and by 1962 it was up to 32,500. General Dynamics made astronautics a full division in 1961. Astronautics managed this rapidly expanding organization as a single project throughout the period.

With the development of different versions of the Atlas and the development of new projects such as the Centaur upper stage and the Azusa tracking system, "priority problems were created in functional line departments, with resultant conflicts over authority and the jeopardizing of performance, scheduling, and cost."[98] Astronautics Division responded to this problem by "utilizing a program control plan called the 'matrix' system which provided a director for each program undertaken by the company." Program directors and department managers resolved priority conflicts.

By 1963 astronautics organized every new major program with the project system using the matrix structure. Atlas Program Director Charles Ames described the organization in the following way:

> Under this system, the program director, . . . is responsible for
> the successful accomplishment of the project, . . . Generally,

personnel working full-time on a project are assigned to the project line organization. The project line activities are organized to fit the specific task. . . . Personnel not assigned to the project line organizations work in functional or "institutional"

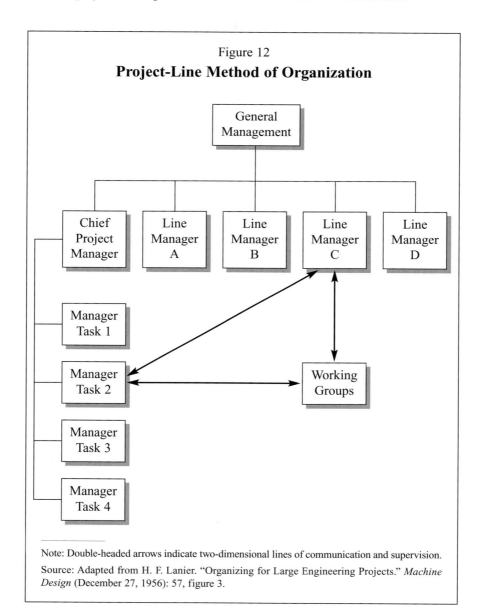

Figure 12
Project-Line Method of Organization

Note: Double-headed arrows indicate two-dimensional lines of communication and supervision.

Source: Adapted from H. F. Lanier. "Organizing for Large Engineering Projects." *Machine Design* (December 27, 1956): 57, figure 3.

departments. Institutional engineering maintains strong scientific and applied research groups as well as preliminary design and systems analysis groups.[99]

Contractors in the aircraft industry found themselves working on several large military projects. Required by the military and prompted through their own complex projects to institute project management, the contractors fit the weapon system concept and its project management into their organizations through the creation of matrix management. Matrix management provided companies with means to move engineers across projects while maintaining disciplinary expertise. It became the industry standard by the 1960s.

On military projects, industry was used to close interactions with government officials. By the early 1950s, the military had checkpoints and approvals at various stages of aircraft development projects. Approvals included those for engineering (design), qualification (meets specifications, proven through test), industry-developed equipment approval (when no government standards exist), installation approval, and preproduction approval (based on "preproduction tests"). Approvals alone were not waivers of contractual requirements but merely "a release to proceed." Along with approval of the overall design, Wright Field engineers and officers also reviewed components, systems, installations, end items, and processes. The Air Force inspected plans for design and production and they inspected the X- and Y-model aircraft.

After those steps had been performed, then came the detailed testing and the design changes that resulted from it. Aircraft service tests included R&D testing; factory acceptance testing; depot and modification center testing; operational suitability testing; and phase testing, which consisted of its own subset of tests including airworthiness and equipment functioning, contractor compliance, design refinement, performance and stability, all-weather, functional development, and operational suitability. The next step was qualification, which was done through inspection, operation, and testing. The Air Force also wanted to verify production through tests and inspections of production lines and the items that came out from them. Finally, the Air Force needed to ensure appropriate environmental testing, which included heat, cold, solar radiation, ice, rain, snow, hail, wind, sand, dust, humidity, fungi, corrosion, shock, acceleration, vibration, low and high pressure, and explosive atmosphere. As these lists make clear, working with the government was not particularly easy and Wright Field was a tough customer.[100]

The missile programs provided their own examples of tight supervision, although because of the novelty of missiles and the speed of their development, the formality of Wright Field's aircraft procedures was somewhat reduced. Despite the management control room device and other information-gathering tools for the Atlas project, Technical Director Charles Terhune was convinced early in the program that Convair was running behind schedule. Eventually he assembled a group of R-W experts and Air Force officers to gather information. They created a set of

**The erector is lowered
away from a Martin
SM–68 Titan ICBM
prior to launching from
Cape Canaveral Missile
Test Annex.**

fifty questions that they sent to Convair to answer. Terhune's group then went to Convair for a week to see what was going on. When Terhune became convinced after a few days that they were getting superficial answers, he had the group stay another week. Convair's president, Joseph McNarney, asked Terhune if he could see the preliminary results of the report, so on the Saturday morning following his group's two-week stay Terhune expected to meet privately with McNarney. Instead he found himself in a room with twenty-five key Convair managers and engineers who were less than happy about what had just happened. Terhune went through his forty findings one by one. Jim Dempsey, the Atlas program manager, told Terhune that after he left the meeting McNarney said, "I don't know if these are true or not, but we're going to go over each one until we have a good answer." Terhune later sent similar "Tiger Teams" to Martin Company because he believed it would have a stimulating effect on Martin's program performance on Titan.[101]

The USAF and the Culture of Innovation

The systems approach to aircraft and missiles had similar implications in industry as in government. Aircraft companies originally organized according to functional groups, with a project leader coordinating the integration of components into the aircraft. The advent of new technologies such as electronics and rocket engines complicated the picture sufficiently that by the early 1950s aircraft companies started to create stronger project organizations, often explicitly including systems engineering. The government–industry team found that ballistic missiles were best organized on a similar pattern, but with some testing and organizational modifications necessary because of their unique technologies. Whereas the government always closely supervised its contractors, the systems approach often made this supervision even closer, with strong project managers and technical consultants like Ramo-Wooldridge looking over the contractor's every task.

Conclusion

The Air Force's ballistic missile programs and the organizations created to support them were primary examples of the Air Force's new project management and weapon system concepts. Using the weapon system concept and supported by scientists, Schriever built the Western Development Division into a powerful organization. His policy of concurrency drove the ballistic missile program at a rapid pace, in keeping with the Soviet threat. Industry reacted by developing matrix management methods that allowed it to manage several projects at the same time. Unfortunately, in their desire to eliminate red tape and bureaucracy Schriever's organization also removed many of the checks necessary to coordinate technically and budget financially for large systems. When combined with the novelty of ballistic missile technologies, the result was a series of test failures compounded by huge cost overruns.

To remedy that situation, the WDD's successor, the Ballistic Missile Division, along with technical direction contractor Ramo-Wooldridge Space Technology Laboratories developed methods to improve missile reliability. These included exhaustive testing, component inspection and tracking, and configuration control to ensure that the design matched the hardware actually launched. The application of these systems engineering concepts eventually led to success, particularly on the Minuteman program where they were consistently applied. A similar pattern of "ad hocracy" followed by more formal methods of engineering coordination was to be found in the same time period on another strategic Air Force program, the defense against nuclear attack, computerized command and control systems.

Notes

1. Presentation, Benjamin N. Bellis, L/Col USAF Ofc DCS/S, "The Requirements for Configuration Management During Concurrency," at AFSC Management Conference, Monterey, Calif, May 2–4, 1962, AFSC, Andrews AFB, Washington, D.C., AFHRA, microfilm 26254, p 5-24-4.

2. Edmund Beard, *Developing the ICBM: A Study in Bureaucratic Politics* (New York: Columbia University Press, 1976), pp 16–20, 120–128.

3. Some good introductions to the Cold War literature include Richard Crockatt, *The Fifty Years War: The US and the Soviet Union in World Politics, 1941–1991* (New York: Routledge, 1995); Walter Lafeber, *America, Russia, and the Cold War 1945–1992,* 7th ed (New York: McGraw-Hill, 1993); Paul Dukes, *The Last Great Game* (New York: St. Martin's Press, 1989); and Stephen J. Whitfield, *The Culture of the Cold War* (Baltimore: Johns Hopkins University Press, 1991).

4. John Lonnquest, "The Face of Atlas: Gen Bernard Schriever and the Development of the Atlas Intercontinental Ballistic Missile 1953–1960," Ph.D. diss, Duke University, 1996, pp 82–97; intvw, Bernard Schriever with author, Mar 4, 1999, USAF HSO.

5. Lonnquest, "Face of Atlas," pp 98–112.

6. Ramo and Dean Wooldridge had just left eccentric billionaire Howard Hughes' company to form one of their own. Jacob Neufeld, *Ballistic Missiles in the USAF 1945–1960* (Washington, D.C.: Ofc of AF Hist, 1990), pp 98–99; and Donald McKenzie, *Inventing Accuracy: A Historical Sociology of Nuclear Missile Guidance* (Cambridge, Mass: MIT Press, 1990), pp 109–110, Lonnquest, "Face of Atlas," pp 113–116.

7. Intvw, Simon Ramo with author, Dec 7, 1999.

8. "Recommendations of the Tea Pot Cmte," Feb 1, 1954, in Neufeld, *Ballistic Missiles,* pp 260–261.

9. Neufeld, *Ballistic Missiles,* p 106; *Space and Missile Systems Organization: A Chronology, 1954–1979* (El Segundo, Calif: Space and Missile Ctr); intvw, Bernard Schriever with author, Mar 25, 1999, USAF HSO.

10. In this context, "in-house" means that Convair elected to build these components inside the company instead of purchasing them from outside. See intvw, Bernard Schriever with author, Apr 13, 1999, USAF HSO.

11. US Cong, House Cmte on Government Ops, *Organization and Management of Missile Progs, Subcmte on Military Operations,* 86th Cong, 1st Sess, 1959, rpt no 1121, p 75.

12. Intvw, Schriever, Mar 25, 1999.

13. Lonnquest, "Face of Atlas," pp 163–170; Beard, *Developing the ICBM,* pp 172–178. Beard believed that the use of R-W was a "fait acccompli" because Asst Secretary of the AF Trevor Gardner personally was promoting the use of R-W in this capacity. Beard interpreted Schriever's defense of R-W as the result of higher-level influences because Gardner was one of Schriever's most important supporters. As we have seen, Schriever had many other reasons to favor the R-W solution, among them his bias to accept the views of the scientists from his previous experience as scientific liaison. Equally important, he accepted the concept of military–civilian partnership for systems integration just as Getting had done at the Radiation Lab. That would have been much more difficult with the large Convair organization than with Ramo and Wooldridge's new company. One problem with prime contractors was that they could bury problems from AF view at lower levels—an action that was far less likely under the associate contractor method. See Jacob Neufeld, ed, *Reflections on R&D in the USAF: An Interview with Gen Bernard A. Schriever and Gens Samuel C. Phillips, Robert T. Marsh, and James H. Doolittle,*

and Dr. Ivan A. Getting, conducted by Dr. Richard Kohn (Washington, D.C.: Ctr for AF Hist, 1993), pp 62–63, for Getting's comments on this, based on his experience at Raytheon, a subcontractor on the B–58 program.

14. Neufeld, *Ballistic Missiles,* p 114; ltr, Ivan A. Getting to author, Jun 18, 1999.

15. Lonnquest, "Face of Atlas," pp 262–263.

16. *Space and Missile Systems Organization,* pp 18–19, 23; Ethel M. De-Haven, *Aerospace: The Evolution of USAF Weapons Acquisition Policy 1945–1961,* vol VI, Hist of Dep Commander for Aerospace Systems 1961, Aug 1962, p 34; memo, Gen Bernard Schriever for the record, subj: Budget Strategy, Oct 2, 1954, item 143, AFHRA, microfilm 35257; memo, Gen Schriever to Col Sheppard and Col Terhune, subj: FY 1955 Reprogramming Actions, Oct 8, 1954, item 140, AFHRA, microfilm 35257.

17. Memo, Gen Schriever to Gen Power, subj: Policy and Procedure for R-W Study Contracts, Oct 2, 1954, item 144, AFHRA, microfilm 35257; memo, Gen Schriever to Col Sheppard, subj: Handling of Contractors, Oct 26, 1954, item 135, AFHRA, microfilm 35257; memo, Gen Schriever to Col Terhune, subj: Competitions for Development Hardware Contracts, Oct 26, 1954, item 133, AFHRA, microfilm 35257; memo, Gen Schriever to Col Terhune, subj: R-W Participation on Evaluation Bds, Jan 24, 1955, item 95, AFHRA, microfilm 35257. Although the evidence for R-W participation in source selection bids is contradictory, it points to R-W contributing information but the AF making final decisions with R-W personnel not present.

18. For the early history of Atlas, see Neufeld, *Ballistic Missiles,* pp 44–50, 68–79; memo Gen Schriever to Gen Power, subj: Airframe Industries vs AF ICBM Management, ca Dec 1954–Jan 1955, item 118, AFHRA, microfilm 35257.

19. See memo, Schriever to Power, Dec 1954–Jan 1955.

20. *Ibid.*

21. Ltr, Lt Gen Thomas Power to All Ctrs, subj: Eligibility of the R-W Corporation and Thompson Products, Inc. to Bid, Feb 24, 1955, AFHRA, microfilm 35258, 168.7171-80; memo, Gen Schriever to Col Ford, subj: General Dynamics—Convair Propaganda Campaign, Mar 18, 1955, item 74, AFHRA, microfilm 35257.

22. Dunn had been hired from JPL, where he had developed the Corporal missile and led the planning efforts for *Sergeant,* on which JPL proposed to use the "system approach." See Stephen B. Johnson, "Insuring the Future: The Development and Diffusion of Systems Management in the American and European Space Progs," Ph.D. diss, University of Minnesota, 1997, chapter 5.

23. Ltr, L.G. Dunn to S. Ramo, subj: Performance of Convair, Mar 22, 1955, AFHRA, microfilm 35258, 168.7171-80; ltr, R. F. Mettler to S. Ramo and L.G. Dunn (STL), subj: Conference at Convair on Friday, Apr 1st, Apr 8, 1955, AFHRA, microfilm 35258, 168.7171-80; ltr, Gen Schriever to Gen Thomas Power, Jun 8, 1955, AFHRA, microfilm 35258, 168.7171-80.

24. Ltr, Col James McCarthy, AF Plant Representative, San Antonio Air Materiel Area, to Gen Schriever, Mar 22, 1955, AFHRA, microfilm 35258, 168.7171-80.

25. Rough draft of "The Atlas Nose Cone," Dec 12, 1956, AFHRA, microfilm 35267.

26. *Space and Missile Systems Organization,* pp 23–25.

27. *Space and Missile Systems Organization,* pp 26–27; Neufeld, *Ballistic Missiles,* pp 119–130.

28. Memo, Brig Gen B. A. Schriever to Col Sheppard, subj: Money, Jan 11, 1955, AFHRA, microfilm 35257; memo, William A. Sheppard, Dep Commander, Prog Management, to Gen Schriever, subj: Money, Jan 12, 1955, AFHRA, microfilm 35257.

29. Neufeld, *Ballistic Missiles,* pp 138–139; Lonnquest, "Face of Atlas," pp 224–244; memo, Gen Schriever to Col Sheppard, subj: Presentation, Apr 1, 1955, AFHRA, microfilm 35257 (I don't know if this presentation was ever created); *Space and Missile Systems Organization,* p 27.

30. Lonnquest, "Face of Atlas," pp 245–248; intvw, Gen Bernard A. Schriever with Dr. Edgar F. Puryear, Jr., USAF Oral Hist Prog, Jun 15 and 29, 1977, AFHRA, K239.0512-1492, p 3; Neufeld, ed, *Reflections on R&D in the USAF,* pp 56–57; *Space and Missile Systems Organization,* pp 28–29; memo, Gen Schriever to Trevor Gardner, subj: Mtg on 6 Sep 1955 to Discuss Management Fund for ICBM Prog, Sep 6, 1955, AFHRA, microfilm 35258, 168.7171-80; DeHaven, *Aerospace,* pp 43–44.

31. Draft memo, Col Ljunggren to Hyde Gillette, subj: Command Relationships—Authority and Responsibility (Weapon Systems—WS-107A), Sep 21, 1955, AFHRA, microfilm 35258, 168. 7171-80; DeHaven, *Aerospace,* pp 43–44.

32. DeHaven, *Aerospace,* pp 51–54.

33. Intvw, Schriever with Puryear, Jun 15 and 29, 1977, p 6; Neufeld, *Ballistic Missiles,* pp 136–137; Gillette rpt, Nov 10, 1955, in Neufeld, *Ballistic Missiles,* p 279; US Cong, House Cmte on Government Ops, *Organization and Management of Missile Progs,* 86th Cong, 1st Sess, rpt no 1121 (Washington, D.C.: GPO, 1959), pp 23–29; DeHaven, *Aerospace,* pp 53–54.

34. "Notes transcribed from an intvw conversation with Col Ray E. Soper, Vice Commander, BSD, on the final day of his service with USAF before retirement," BSD(BEH) NAFB, Cal 92409, 29 Nov 1966, AFHRA, microfilm 30015, p 3.

35. *Ibid.*, p 5.

36. Some historians have questioned whether the Control Room served any real purpose aside from showmanship. My conclusion is that some of the people in the WDD saw that it had a "show" purpose and concluded that is all that it was for. After interviewing Charles Terhune, I am convinced that it served a real purpose in giving the "official" status at any time, and in providing a central information repository. See intvw, Charles Terhune with author, Sep–Oct 1998, USAF HSO. Schriever also believed it served a useful purpose in precisely that way, not simply "for show." See intvw, Schriever, Mar 4, 1999.

37. Lonnquest, "Face of Atlas," pp 258–259, 266–267, 291–297; intvw, Schriever, Mar 4, 1999. Schriever believed that the idea for Black Saturday was his own because his philosophy of management was to dig out the problems and not spend time on the status of things that were going well. He did not recall seeing this practice in any other programs at that time. It also seems that Schriever's group soon began to use computers for the Management Control System, with R-W. That would make it one of the earliest applications of computers to management communication.

38. Lonnquest, "Face of Atlas," pp 300–302.

39. Harvey Sapolsky found a similar phenomenon in the use of the PERT for the Navy's Polaris program; see Harvey M. Sapolsky, *The Polaris System Development, Bureaucratic and Programmatic Success in Government* (Cambridge: Harvard University Press, 1972).

40. Lonnquest, "Face of Atlas," pp 300–311.

41. Ethel M. DeHaven, *AMC Participation in the AF Missiles Prog through Dec 1957* (Inglewood: Ofc of Information Services, AMC Ballistic Missile Ofc), AFHRA, roll 2312, pp 177–181; US Cong, *Organization and Management of Missile Progs,* pp 30–32; memo, Gen Schriever to Col Terhune, Jan 24, 1955.

42. Memo, Gen Schriever to Col Terhune, subj: Procedures Governing Development of Facilities, Mar 17, 1955, item 76, AFHRA, microfilm 35257; De-Haven, *AMC Participation,* pp 177–181

43. DeHaven, *AMC Participation,* pp 290–291.

44. Memo, Col Terhune to Gen Schriever, subj: Continuous Monitoring of USAF Atlas Development Contracts, Dec 21, 1954, item 109, AFHRA, microfilm 35257; memo, Col Sheppard to Gen Schriever, subj: (unclassified) WDD Fiscal Management w/Chart, Jan 24, 1955, item 108, AFHRA, microfilm 35257; memo, William A. Sheppard, Dep Commander for Prog Management, to staff, subj: Contractor Supervision, Jan 17, 1955.

45. DeHaven, *AMC Participation,* p 171.

46. *Ibid.*, pp 182–183.

47. *Ibid.*, p 183.

48. *Ibid.*, pp 185–187.

49. Intvw, Terhune, Sep–Oct, 1998; memo, Gen Schriever to Gen Power, subj: R-W Contract Administration and Study Subcontracts, Nov 1, 1954, item 130, AFHRA, microfilm 35257.

50. Memo, Gen Schriever to Gen Power, Nov 1, 1954.

51. DeHaven, *AMC Participation,* pp 185–186; memo, Gen Schriever to Dr. Ramo, subj: R-W Organization and Procedures, Apr 20, 1955, item 59, AFHRA, microfilm 35257.

52. *Research and Development in the USAF* [Ridenour rpt], SAB, Sep 21, 1949, AFHRA, 168.1511-1, pp IX-1, 2.

53. "Gen Doolittle's Summ," Jan 15, 1952, MDP, AFHRA, 168.7265-235.

54. Presentation, Col M. C. Demler to CofS and Comptroller, WADC, on Management Practices, Feb 15, 1952, AFHRA, 168.7265-235.

55. Presentation, Maj Gen Donald L. Putt, Commanding Gen, WADC, to Chiefs of WADC Labs, "Organizational Philosophy of the AF," Apr 2, 1952, AFHRA, 168.7265-235

56. Memo, Donald L. Putt, Vice Commander, to Commanding Gen, WADC, subj: General Policy Guidance on Use of Single Prime Contractor for Development of a Complete Weapon System, Dec 8, 1952, MDP, AFHRA, 168.7265-237.

57. Memo, S. R. Harris, CofS, to Commanding Gen, WADC, subj: Policy on Testing of Guided Missiles, ca Dec 1952, MDP, AFHRA, 168.7265-237; Booklet prepared by Directorate of Weapon Systems Op, WADC, ARDC, *Development Life Cycle of a Weapon System,* ca Dec 1954, MDP, AFHRA, 168.7265-237.

58. Ltr with attached charts, Simon Ramo to Gen Schriever and others, subj: The R-W Corporation's Guided Missile Research Div Organization, Apr 26, 1955, AFHRA, microfilm 35258, 168.7171-80.

59. *Ibid.*

60. US Cong, Hearings before the House Subcmte for Special Investigations of the Cmte on Armed Services, 86th Cong, 1st Sess, *Weapons System Management and Team System Concept in Government Contracting* (Washington, D.C.: GPO, 1959), pp 202–204; Robert J. Art, *The TFX Decision: McNamara and the Military* (Boston: Little, Brown and Company, 1968), pp 84–85; intvw, Col Edward N. Hall with Jack Neufeld, USAF Oral Hist Prog, Jul 11, 1989, AFHRA, K239.0512-1820, p 6; fourth rough draft, "The Atlas Guidance System," Dec 27, 1956, RW-EO(I)-011-56-A, item 25, AFHRA, microfilm 35267; 3d rough draft, R-W's Contribution to the Titan Missile, Dec 28, 1956, RW-EO(I)-011-56-C, item 24, AFHRA, microfilm 35267; "The IRBM Preliminary Analysis and Design Specification," Jan 17, 1957, RW-EO(I)-011-56-D, item 21 AFHRA, microfilm 35267.

61. Excerpt from Suppl Agreement #9, Oct 24, 1956, AF18(600)-1190, item 31, AFHRA, microfilm 35267.

62. Memo, Lt Col Bogert for the record, subj: Information Submitted to Col Covemar in Gen Irvine's Ofc Regarding R-W, Oct 26, 1956, item 30, AFHRA, microfilm 35267; memo, Gen Schriever to CMDR/AFMTC, subj: Management Requirements (w/Attachment—Draft of Integration of WDD into Normal Structure of ARDC), Nov 6, 1956, item 29, AFHRA, microfilm 35267.

63. HQ ARDC, *Policies and Procedures Governing Approval of AF Equip-*

ment, Baltimore, Aug 1, 1952, MDP, AFHRA, 168.7265-237; WADC Regulation 80-25, "Research and Development Programming Procedures," May 25, 1953, AFHRA, MDP, 168.7265-237; Directorate of Weapon Systems, *Development Life Cycle of a Weapon System.*

64. Ltr, Gen Schriever to Maj Franzel (RDGB), subj: Test Philosophy for Guided Missiles, Requesting Gen Powers' Signature, Feb 7, 1955, item 104, AFHRA, microfilm 35257. That letter includes a memo to the CofS, HQ USAF, from WDD HQ ARDC, subj: Test Philosophy for Guided Missiles.

65. *Ibid.*; "The Role of "R-W in the Ballistic Missile Testing Prog," Jan 1957, RW-EO(I)-011-56-E, item 23, AFHRA, microfilm 35267.

66. Symposium on Guided Missile Reliability, Nov 2–4, 1955, sponsored by the DOD under auspices of ARDC, AF-WP-O-21, Sep 1956, 56RDZ-12531, Wright-Patterson AFB, Ohio, AFHRA, microfilm 26254.

67. Presentation, Maj Ernst P. Luke, Ernst Lange, Holloman Air Development Ctr, "Weapon System Evaluation and Reliability," at Symposium on Guided Missile Reliability, pp 169–170.

68. Neufeld, *Ballistic Missiles,* pp 178–179, 205–206; Atlas prog information prepared for Stennis Subcmte, May 20, 1961, AFHRA, K243.012-42.

69. Intvw, Hall, Jul 11, 1989, pp 5–7.

70. Neufeld, *Ballistic Missiles,* pp 217–220; "Excerpts from Task Force's Rpt to Kennedy on US Position in Space Race," *New York Times,* Jan 12, 1961. For details about Atlas failures, causes, and responses, see Problems of the Atlas Prog: Status—Corrective Action, Apr 26, 1965, AFHRA, box "J6 and J7," no file # or name.

71. Ltr, Col Soper to WDG, subj: (U) Missile Reliability, Accuracy, Apr 1, 1960, AFHRA, box "J3 thru J5," file "ICBM-1-19, Ops Reliability [1959]."

72. Ltr, Gen Terhune to Gen Ritland, subj: AFFTC Resources Application to Rocket Development, Apr 19, 1960,

AFHRA, box "J3 thru J5," file "ICBM-1-19, Ops Reliability [1959]."

73. Telex, from SAC Offutt to HQ USAF, Oct 19, 1960, AFHRA, box "J3 thru J5," file "ICBM-1-19, Ops Reliability [1959]."

74. Neufeld, *Ballistic Missiles,* pp 217–219.

75. DeHaven, *AMC Participation,* pp 302–304.

76. Atlas prog information for Stennis Subcmte, "Reliability Prog" section.

77. *Ibid.*

78. Bellis, "Requirements for Configuration Management."

79. Lysle A. Wood, "Configuration Management," AFSC Management Conference, Monterey, Calif, May 2–4, 1962, AFSC, Andrews AFB, Washington, D.C., AFHRA, microfilm 26254, pp 5-23-1–4; statement, Clyde Skeen, Asst Gen Manager, Boeing Airplane Company, Apr 30, 1959, in US Cong, Hearings before the House Subcmte for Special Investigations of the Cmte on Armed Services, 86th Cong, 1st Sess, *Weapons System Management and Team System Concept in Government Contracting* (Washington, D.C.: GPO, 1959), pp 201–207.

80. Paper, author unknown, subj: Weapons Systems Engineering on the USAF Ballistic Missile Progs—The R-W Corporation Responsibility, ca 1957, item 20, AFHRA, microfilm 35267.

81. Bellis, "Requirements for Configuration Management"; Wood, "Configuration Management."

82. Atlas prog information prepared for Stennis Subcmte, May 20, 1961, AFHRA, K243.012-42, "Reliability Prog" section.

83. HQ AFSC, *AFSC Manual, Systems Management, AFSCM 375-1: Configuration Management During the Acquisition Phase,* Jun 1, 1962, LOC/SPP, box 18.

84. *SRA Orientation Guide* (San Bernardino: Space Tech Labs, Inc., Thompson Ramo Wooldridge, Inc., ca 1960), prepared for the Minuteman Prog Ofc of AFSC, LOC/SPP, box 18.

85. Lonnquest, "Face of Atlas," p 308.

86. See intvw, Terhune, Sep–Oct, 1998.

87. Intvw, Hall, Jul 11, 1989; see also intvw, Terhune, Sep–Oct, 1998; intvw, Schriever, Apr 13, 1999. Hall's personality is very clear from his intvw with Neufeld. He spoke his mind, and valued technical knowledge above all else. Tact was not his strong suit, which led Schriever and Terhune to put him in positions where his abrasiveness would not be a detriment.

88. See Irving Brinton Holley, Jr., *Buying Aircraft: Matériel Procurement for the Army AFs,* vol VII of Stetson Conn, ed, *US Army in World War II* (Washington, D.C.: Ofc of the Chief of Military Hist, Dept of the Army, 1964); see also this book, chapter 3.

89. William B. Harwood, *Raise Heaven and Earth: The Story of Martin Marietta People and Their Pioneering Achievements* (New York: Simon & Schuster, 1993), pp 253, 278; Edward G. Uhl, "Applying the Systems Method to Air Weapons Development," *Aviation Age* (Feb 1954), pp 20–23; William B. Bergen, "New Management Approach at Martin," *Aviation Age* 20 (Jun 1954), pp 39–47. Bergen later became manager of North American's Apollo program.

90. Glenn E. Bugos, "Manufacturing Certainty: Testing and Prog Management for the F–4 Phantom II," *Social Studies of Science* 23 (1993), pp 271–275.

91. Keith Davis, "The Role of Project Management in Scientific Manufacturing," *IRE Transactions on Engineering Management,* vol EM-9, no 3 (Sep 1962), pp 109–110.

92. H. F. Lanier, "Organizing for Large Engineering Projects," *Machine Design* (Dec 27, 1956), p 54.

93. *Ibid.*, p 55.

94. *Ibid.*, p 56.

95. *Ibid.*, p 57.

96. *Ibid.*

97. Charles S. Ames, "The Atlas Prog at General Dynamics/Astronautics," in Fremont Kast and James Rosenzweig, eds, *Science, Technology, and Management* (New York: McGraw-Hill, 1963), pp 199–201.

98. Ames, "The Atlas Prog," pp 201–203.

99. *Ibid.*, p 204.

100. HQ ARDC, *Policies and Procedures.*

101. See intvw, Terhune, Sep–Oct, 1998. According to Schriever, the "Tiger Team" idea was never formalized but was used occasionally, as Terhune confirmed; intvw, Schriever, Mar 4 and 25, 1999.

Chapter 4

To Command and Control

The work now starting on SAGE master programs will require for-
malities that have not been necessary in the past. We are no longer
preparing programs for "our" system.

C. Robert Wieser, March 1955[1]

The Soviet threat of the late 1940s through the 1950s centered on the ability
of the USSR to recreate U.S. successes in the development of nuclear fission
and, later, fusion weapons. In response to this threat, Air Force leaders believed
above all else that the best defense was a good offense. The experience of World
War II taught them that no matter how good air defenses might be, some bombers
always broke through to attack their targets. Therefore they placed their bets on
offensive capability, first with strategic bombers and then, grudgingly, with ballistic missiles. They realized also, however, that they needed some kind of air defense or at least an early warning capability, if only to protect their bombers.[2]

Because the Soviets were developing long-range bombers capable of attack-
ing targets in the continental United States, Air Force leaders ultimately decided
to begin developing a continental air defense system in cooperation with their
counterparts in Canada. As Soviet jet bombers and then ballistic missiles became
a threat and the warning time available to launch a nuclear counterstrike shrank
from hours to minutes, maximizing those precious few minutes of early warning
became even more critical. Nuclear war presented Air Force officers with stark
options that had to be executed with little or no deliberation. Operational commanders had to determine in a matter of minutes whether the appearance of unknown aircraft or missiles on a warning system device constituted a nuclear strike
or rather were phenomena related to atmospheric conditions or technical problems in the warning system itself. Based on that assessment, commanders then
had a minute or two in which to order a nuclear counterstrike with bombers and

later with missiles. A mistake either way—launching a strike erroneously or failing to launch the counterstrike quickly in case of an actual attack—would lead to the destruction of the United States. In such circumstances, Air Force leaders wanted every extra second to make the proper decisions. They also demanded that their information be reliable and then that their commands be executed rapidly. To maximize decision time, the time needed for data collection, transfer, and interpretation had to be minimized.

In World War II, people in one location had collected data and used telephones or radios to transfer information to a centralized air defense control center where other people placed the relevant information on a situation map. Others there had interpreted the data and the commander then decided what actions to take—whether to launch interceptors, to warn potential targets, or to declare the information erroneous or irrelevant (for example, when the aircraft detected were either commercial flights or friendly aircraft on expected flight paths). During that war, air battles extended over several hours and thus left some time for deliberation and assessment. In the jet and missile age, however, that time shrank to mere minutes and pushed Air Force leaders to search for ways to speed up all of the processes.

Just as technology in the form of jet and rocket engines caused the problem, technology in the shape of radar and computers provided possible solutions. Although commanders in no way wanted machines to replace their decision functions, they were perfectly happy to have computers and radar replace people elsewhere in the chain of command if doing so made the relevant processes faster and more reliable. Radar extended the senses of the military and computers promised to speed their reactions. These technologies were potentially attractive solutions to their critical need for speed and accuracy.

Although radar was a well-developed technology by 1950, computing was not. Several computers were under development by this time and, fortuitously, one known as the Whirlwind computer at MIT was being developed with so-called real-time applications in mind. In computer jargon, "real-time" refers to the capability to perform operations as fast as or faster than necessary in real life. For example, for a computer to simulate the flight of an aircraft so that a pilot could use it for training, it would have to make calculations of the simulated maneuvers as fast as or faster than the maneuvers themselves. In the air defense application, it did little good if the calculation of an aircraft's position took several hours to complete. The computer had to calculate radar information and aircraft positions and place the data on a cathode-ray-tube screen as fast as or faster than the sensed input data. Whirlwind had the processing speeds necessary to complete these calculations quickly enough and was predicted to have the reliability sufficient to operate continuously.

The Air Force adopted Whirlwind, which became the core of the continental air defense system and the prototype for other systems to give early warning against missile attack, to track space objects, and to allow commanders to collect data and issue commands rapidly. These command and control systems proliferated just as ballistic missiles had done and they posed similar organizational problems. Just as in the case of ballistic missiles, the Air Force created new organizations and new processes to manage these complex technologies. It called on civilian scientists and engineers to help create and organize programs and machinery that again would be produced by private industry. In this process the newly formed computer industry was shaped profoundly by the Air Force as it tried to bring about continental warning and defense.

The Navy's Problem Child

What would eventually become the core of the United States' air defenses started out as a World War II Navy project to develop an all-purpose aircraft training simulator. Captain Luis de Florez, the technical director of the Special Devices Division of the Bureau of Aeronautics, believed that the time had come to develop a general-purpose aircraft simulator that could act like a not-yet-built aircraft. This device would allow advanced engineering analysis and could be used to train Navy pilots. He contracted for the task with the Servomechanisms Laboratory at MIT because he and the laboratory expected the task to involve the development of sophisticated feedback control systems using servomechanisms. Funding for the aircraft stability and control analyzer (ASCA) began in 1944, and Gordon Brown, the director of the Servomechanisms Laboratory, placed Assistant Director Jay W. Forrester in charge of the project.[3]

Forrester was a meticulous, hard-driving electrical engineer who had come to MIT in 1939 as a graduate student and research assistant. During World War II he got involved with the many war-related projects in the Servomechanisms Laboratory. Soon after taking the lead on the ASCA project, Forrester brought on Robert Everett, an electrical engineer who had come to MIT as a graduate student in 1942. Forrester quickly realized that the most difficult portion of the job would be developing the general-purpose computing device at the heart of the flight trainer. It would have to operate very rapidly and reliably while maintaining a high level of accuracy in its computations, and be flexible enough to switch among simulations of different aircraft. Forrester initially pursued the development of an analog computer to perform these tasks, but by the fall of 1945 he had begun to toy with the idea of using a general-purpose digital device similar to the ENIAC, which had been well publicized at the end of WWII.[4]

By February 1946 the Navy accepted Forrester's proposal that the trainer's core should be a digital computing device, and it accepted the additional costs and an extended schedule. At that time the Navy dubbed the project Whirlwind, the name ultimately given to the computer. Forrester projected the system would be completed in June 1948 at a total cost of nearly $1,200,000. Soon the MIT group was enmeshed in the details of designing a digital computer that rapidly could calculate the equations necessary to simulate flight. Forrester recognized that such a machine would be useful for many other applications and in his mind that made its development all the more important. Unfortunately for him the significance that the Navy assigned to the project soon would decrease dramatically and that would lead to conflicts among Forrester, MIT's management, and the Navy.[5]

The Navy's assessment of Project Whirlwind changed because of significant changes in the Navy's organization and consequently the Navy personnel supervising the project. The Special Devices Division became part of the newly created Office of Naval Research in late 1946, and after a short period spent assessing the Navy's various research projects, ONR managers decided to evaluate Whirlwind more meticulously.

The immediate cause of the ONR's concern was Forrester's semiannual report of January 1947. Forrester proposed developing a "pre-prototype" computer to be followed by the actual prototype machine. In his January report, he stated that the start of the prototype development would have to be delayed six months to January of 1948. In addition, technical problems with the storage-tube memory were such that Forrester could make no promises about its probability of success. Although these statements were not unexpected or disturbing to the Special Devices Division, the ONR considered them differently. The Whirlwind project stood out from other computer projects that the ONR was funding (such as the Institute for Advanced Studies computer at Princeton) because its cost was substantially higher. Whirlwind's full cost eventually would be estimated at around $3,000,000, whereas ONR managers projected that none of the other computers it funded would cost more than $650,000. Whirlwind's history was worrisome to ONR's managers. It had started out as a simulator–trainer project, had narrowed its focus to a digital computer, and now appeared to be hedging its bets even on that, all for costs far higher than any of the ONR's other computer projects.[6]

This assessment led to a review of Whirlwind by Warren Weaver, the head of the Naval Research Advisory Committee. He found that, although the personnel seemed hard at work doing a reasonable job with the physics and engineering, they might be lacking expertise in the mathematical aspects of the design. This did not sit well with Dr. Mina Rees, the head of the ONR's mathematics branch. Whirlwind's high costs and apparent lack of progress troubled Rees and the report did not assuage her concern. Forrester responded to Weaver's critiques with a twenty-two-volume summary of Whirlwind's development since 1944.[7]

Furthermore, the Special Devices Center (as the Special Devices Division was renamed) requested that Forrester subcontract for the hardware to speed development. This request did not help matters because the design of Whirlwind was not yet completed. Forrester responded by subcontracting with Sylvania to manufacture the computer hardware. With the design still fluid, however, there were substantial overruns. Forrester's group devoted itself to hammering out a design and to checking and double-checking the design drawings given to Sylvania.

The several design changes they made after delivering drawings to Sylvania further ran up costs. Sylvania's overhead rate was 137 percent and that led to higher prices for subcontracting than if the work had been done at MIT.[8]

Still concerned about the project, the ONR sponsored another review in November 1947, this time by Columbia University mathematician Francis Murray. He reported some weakness in the mathematical area, but he stated that Whirlwind's high costs might be justified because of the emphasis on engineering necessary for its simulation application. In other words, this computer would have to be more reliable than comparable computers (such as Princeton's machine) that were designed only for mathematical computations. That idea apparently resulted from Weaver's earlier visit because he had suggested to Forrester that computer development work was "becoming more appropriate for an engineering organization" as opposed to a research group. John von Neumann, the famous mathematician in charge of the Princeton project, rebutted that opinion, stating that his machine would have to pay equal attention to engineering. He viewed the differences between his machine and Forrester's as the product of different personal judgments regarding the best technologies and approaches, not of different intended applications or "deep differences of principle."[9]

The same military budget cuts that led to the cancellation of Convair's ICBM efforts in 1948 led to budget pressures against Whirlwind. With the project consuming a substantial portion of the ONR's budget, the ONR became convinced that it had to rein in Forrester's extravagance. Things drew to a head in the fall of 1948. As ONR debated means to control the project, Forrester and the Special Devices Center requested $1,831,583 for the fifteen-month period from July 1, 1948, to September 30, 1949. This request went far beyond previous estimates and wreaked havoc with the ONR's budget. The resulting furor led to a confrontation between the administration of MIT and the ONR.[10]

MIT's top administrators had already become concerned about the deteriorating relationship between the project and the ONR, and they had assigned a member of MIT's electrical engineering committee and a member of the Harvard University physics department to evaluate the project in the spring and summer of 1948. Their report had been quite favorable, stating that "accomplishments . . . give every promise of providing within the scheduled date a successful computer at speeds hitherto unrealized." By September 1948 the chief of naval research contacted MIT president Karl Compton to try to bridge the gap between the project's goals and the Navy's wish to cut costs. Forrester and his team met with Compton to describe the project and the promise of digital computing for the future of the United States and for MIT. Compton requested a detailed report from Forrester. Armed with that report, Compton then met with the ONR. The meeting on September 22, 1948, ended in a compromise, with the Navy eventually agreeing to supply $1,200,000 through June 30, 1949. However, the Navy refused to budge from its fiscal year 1949 allocation of $750,000 for July 1, 1949, to June 30, 1950.[11]

The ONR's efforts to gain control over the project resulted in the transfer of the project from its supporters in the Special Devices Center to its critics in the mathematics division. When the transfer was complete in February 1949 Forrester could look forward to continuing close supervision and further critical reviews. Rees asked John von Neumann to visit the laboratory in February 1948, and later she requested that two researchers from the Bureau of Standards consider "in some detail with the Project Whirlwind Staff the nature of engineering problems of computer design and the successive stages of development leading to the final product." In the spring of 1949 Rees arranged for a review team headed by Dr. Harry Nyquist, a respected electrical engineer at Bell Telephone Laboratories. He and his committee were favorably impressed by the project, but later that year he was involved with a second review that was far less optimistic.[12]

In the summer of 1949 the Research and Development Board created the Ad Hoc Panel on Electronic Digital Computers under its Committee on Basic Physical Sciences. The panel was to review critically all computer projects then under way in the armed services. Their report of December 1, 1949, raised a storm of protest not only from Forrester but also from many other computer developers. Finding that Whirlwind lacked a "suitable end use," the panel recommended that "further expenditure . . . should be stopped" unless a suitable use could be found. Although Forrester and other computer developers attacked the results of the panel, the damage had been done. Whirlwind was the ideal target for saving money because of its high costs and lack of end use.[13]

In March 1950 further meetings between the ONR and the MIT administration led to an agreement that the ONR would provide $280,000 more for the fiscal year starting July 1950, but that would be the end of the Navy's commitment. Whirlwind needed more funding before it could be completed and another suitor arrived to save the project—the Air Force.[14]

The Air Force Reaps the Whirlwind

The impetus for the Air Force's involvement with Whirlwind came from the efforts of MIT physicist George E. Valley. Despite Valley's presence at the same school as Forrester, his discovery and selection of Whirlwind for Air Force purposes was serendipitous. Valley's concern was not computing but rather the poor state of U.S. air defense following the successful test of a Soviet atomic weapon in the fall of 1949. A veteran of MIT's Radiation Laboratory in World War II, he had been involved with bombing radar rather than air defense. Following the war he worked on cosmic ray research, but after the Soviet atom bomb test he devoted his efforts to improving U.S. air defenses through his membership on the SAB's Electronics Panel. Valley's inspiration came in part after he purchased a new house on a hill facing Boston and realized that in the case of a Soviet strike his home would have little protection.[15]

MIT physicist and Radiation Laboratory veteran George E. Valley got the Air Force involved in the Whirlwind project. He believed that the major technical air defense problem to be solved was computing the positions of aircraft quickly enough to predict their future positions.

Valley soon found that air defense was in a dismal state. Poking around the Boston area for information, he visited a local radar station and the Air Force Cambridge Research Laboratory (AFCRL), an Air Force facility established across from the MIT campus to lure Radiation Laboratory veterans like Valley into Air Force tasks. AFCRL director John Marchetti confirmed the dilapidated state of air defense. He also showed Valley a number of promising new technologies, including means to pass radar data through the telephone network using what would become known as a modem, and an early version of a "light gun," a device that could select an element in a cathode-ray tube (like a regular television screen) by pointing at the location and pushing a button. Last, Valley contacted the SAB to request some reports and to speak with its chairman, Theodore von Kármán. Von Kármán asked him to write a brief summary of his investigation and ideas.[16]

Valley sent a three-page summary to von Kármán on November 8, 1949, proposing the formation of an Air Defense Committee in the Boston area, under the purview of the SAB, to further investigate the problem of air defense and possible technical solutions to it. The letter found its way to Gen. Muir Fairchild, Air Force vice chief of staff. By December 15 Fairchild requested that Valley be the chair of the Air Defense Systems Engineering Committee (ADSEC).

Valley believed that the major technical problem to be solved was computing the positions of aircraft quickly enough to predict their future positions. He and Marchetti reasoned that using one of the new digital computers might solve the problem.[17] The ADSEC met every Friday at the AFCRL. After reviewing the

MIT's Jerome B. Wiesner (shown here) connected George Valley with the Whirlwind computer sitting "up for grabs" on the MIT campus.

various computing systems under development at the time, Valley could find no computing group interested in the air defense application. In a chance meeting with Jerome Wiesner in a hall at MIT in January 1950, however, Valley learned that there was a computer "up for grabs, right there on the MIT campus." Valley went to see Forrester while Marchetti checked with his ONR counterparts about Whirlwind. Both men heard the negative criticisms of Whirlwind but Valley concluded that most of the criticisms were unfounded. For the air defense problem, which required rapid computation and high reliability, Whirlwind's speed and "elephantine construction" were significant advantages. And Whirlwind had another significant advantage: it was available.

The Air Force's Watson Laboratory had a small contract with Forrester to evaluate the application of Whirlwind to air traffic control and Marchetti arranged for Valley to take control of that contract. Valley also was encouraged that the group working on Whirlwind had learned a few things about aviation and that neither Forrester nor Everett "seemed to think he knew all the answers." By January 1950 Valley saw that the computer was running some simple physics problems, demonstrating its initial operational capability. This convinced him and Whirlwind became the prototype computer for his proposed air defense system.[18]

In March 1950 Valley attended a meeting with the ONR and the MIT administration concerning funds for Whirlwind, and in principle he promised $500,000 from the Air Force to complement $280,000 from the ONR for fiscal year 1951. Valley officially introduced the Air Force to the idea of using a digital computer in his progress report in May 1950. By November 1950 the Air Force added $480,000 in funding to continue Whirlwind's development. Forrester reoriented the project to air defense and by September 1950 demonstrated that Whirlwind could read radar data from telephone lines and display them on a cathode-ray tube.[19]

At that point, Louis Ridenour and Ivan Getting, Valley's former compatriots at the Radiation Laboratory, intervened to keep Valley's services and to involve MIT once again in defending the nation. At an earlier point MIT had turned down some Navy work, not wishing to perform classified military research. But with the successful Soviet bomb test of the previous year and the Korean War under way, Ridenour and Getting thought the MIT administration might be coaxed to recreate something like the Radiation Laboratory. Ridenour, then the scientific

Jay Forrester and Norman Taylor (shown left) discuss the computer's components while Gus O'Brien, Charles Corderman, and Norm Daggert test Whirlwind equipment among the racks in the computer room at MIT's Barta Building.

adviser to Air Force DCS/D Gordon Saville, and Getting, the assistant for evaluation (soon to be development planning) in the DCS/D office, traveled with Saville to visit MIT president James Killian in the fall of 1950. They persuaded Killian that MIT once again should take up the cause. Ridenour followed up his trip by writing a memorandum on November 20 to propose that MIT take a more formal role in the air defense effort, which would be based in the Cambridge area to support Valley's committee and the Air Force Cambridge Research Laboratory. On December 15 Air Force Chief of Staff Vandenberg made a formal request to MIT, and on January 30, 1951, the Air Force issued a letter contract to MIT to investigate the development of an air defense system.[20]

At the instigation of other powerful scientists who had not been involved, the MIT administration briefly balked but eventually agreed to take a more active role. The so-called ZORC group, named for its main actors Jerrold Zacharias, J. Robert Oppenheimer, Isador I. Rabi, and Charles Lauritsen, supported strong air defenses as part (some claimed) of a larger moral struggle against the development of hydrogen weapons. On their recommendation MIT's involvement soon required two further conditions. First, the proposed laboratory would have to be a triservice arrangement so as not to threaten MIT's Navy

MIT president James R. Killian. He agreed that MIT should aid the nation's need for air defense while ensuring extraordinary authority for Lincoln Laboratory.

contacts and contracts. Second, the problem would have to be studied in more detail beyond Valley's committee. These conditions led to a second detailed investigation by a broader base of experts influenced by Zacharias' group. The new study, known as Project Charles and run from January to July 1951, featured another demonstration by Forrester's group, this time showing that Whirlwind could track an aircraft and display the results on a cathode-ray tube, and that an operator could order a fighter to intercept the mock-bomber. On July 26, 1951, the three services signed onto the laboratory project, and Project Lincoln was born under MIT management. In February 1952 MIT signed the formal contract which was to run until January 1, 1959.[21]

Lincoln Laboratory had support at the highest levels of the Air Force but it also had a number of skeptics, both inside and outside of the service. Its strongest support came from high-level officers who remembered the stellar contributions of the World War II Radiation Laboratory and the competence of MIT's participating faculty. In the new postwar emergency, members of the Air Staff resurrected the experiences of World War II as a model for operating the new laboratory. They stated that "the proposed MIT Air Defense Laboratory is the first activity intended to mobilize scientists and engineers for important USAF research and development work. . . ." Because of better working relationships with scientists and engineers after World War II, they believed that "our success in establishing these university-operated laboratories may well obviate the necessity for a new NDRC–OSRD." If successful in operating these new civilian laboratories, the Air Force would accrue the benefits and be better placed to direct the research in ways that benefited its service branch.

To be successful, Air Staff members recognized that they would need to give the civilians substantial freedom and flexibility. Ivan Getting developed the management concept for the new laboratory with the idea of maximizing its freedom and flexibility.[22] In a memo to the commanders of AMC and ARDC, DCS/M Lt. Gen. Kenneth B. Wolfe laid down the ground rules:

    a.   Maximum latitude and flexibility in the interpretation of procurement regulations in writing contracts to cover work done in these laboratories;

    b.   Fullest availability to each laboratory of the problems, developments, and plans of the military departments in the laboratory's field of interest;

    c.   Reasonable freedom for each laboratory to initiate and manage specific projects in its program area. To the maximum extent possible, unnecessary duplication will be prevented primarily by insuring that each laboratory is well informed about work already being done in its field.[23]

Upon the transition to Project Lincoln and Lincoln Laboratory, a triservice steering committee known as the Lincoln Advisory Committee provided official

Early computer graphics displayed a simple message when the Whirlwind computer was featured on Edward R. Murrow's television show "See It Now" in the 1950s.

127

military management. Although each service had one official representative on the committee, the Air Force's predominant interest and financial support of the laboratory meant that its voice generally prevailed. The AFCRL, now renamed the Air Force Cambridge Research Center, provided local support. At Lincoln, AFCRC established a small liaison office with a colonel, a field-grade officer, and a company-grade officer, and it eventually became the Lincoln project office. Personnel there found themselves in a difficult situation for although they had responsibility for the project, Lincoln Laboratory held the real power. The supervisors could do little without the approval of their erstwhile subordinates.[24]

Through AFCRC, Air Research and Development Command coordinated Lincoln's requirements with the rest of its R&D programs. Despite the "emergency" that required mobilizing scientists and engineers on the World War II model, Air Force leaders did not consider the program to be as critical as was its offensive deterrent through SAC. Despite Valley's request to increase the project's priority for scarce materials and resources, Brig. Gen. Donald Yates, the director of R&D in the office of the DCS/D, informed Project Lincoln's leaders that top priority would not be forthcoming. Because the Air Force deputy of operations set these ratings "strictly in accordance with war plans," and the procedures involved many complex considerations, a priority increase would be very difficult. Yates believed, in general, that ARDC's ratings were "proper in its relationship to the ratings of other Air Force activities." Yates realized that this would impose difficulties "but with a few probable exceptions, not insurmountable ones." It meant that Lincoln personnel would have to provide individual justifications for components and to coordinate with other Air Force commands that would be adversely affected. Despite attempts to elevate the laboratory's priority over the next year, the Air Staff refused to increase it.[25]

With the establishment of the Lincoln Laboratory, the Air Force committed itself, with some hesitation, to developing new technologies for air defense. Its leaders believed that air defense was a generally dubious proposition and were unwilling to commit their top priority and massive funds to the project, at least until it proved its utility. They saw the potential of new technologies to improve air defense but were unsure about the use of the Navy's cast-off digital electronic computer. Project Lincoln as a whole and the Whirlwind team in particular would have to prove themselves worthy of such attention and, with the technical and organizational competence they had built since World War II, Forrester's team would meet that challenge.

Organizing a Controversial Computer Project

The aircraft stability and control analyzer began as a typical World War II R&D project. Serving the Navy's needs for a trainer while also whetting MIT's interest in challenging technical projects for its students, ASCA's informal project structure also seemed typical of a military-sponsored academic project. The informality was deceptive, however, for behind the academic façade Jay Forrester ran a very tight ship. In many ways his organization resembled corporate America more than academia.

Forrester kept close watch over his organization through both formal and informal means. Informally, he often wandered about the laboratory, checking in with the graduate students and full-time personnel, questioning them about their work and its progress. He was very tough and often bluntly set them off in a new direction if he believed their efforts to be ill focused. As one of the engineers later recalled, "Forrester would come into the lab and tear everything apart, and Bob [Everett] would come along and put it back together again." Another engineer added, "Tear you apart, you mean." No one questioned Forrester's ability or his dedication to the project. He also held weekly "Friday afternoon tea" sessions in his office at which invited personnel shared more detailed expositions of their work with the division heads and others. Although informal in appearance they were serious in intent, as was Forrester himself.[26]

From the start of the project, Forrester required extensive documentation. By early 1946 he had developed six different series of documents, each with a specific purpose and distribution. These included administrative memoranda ("A"), conference notes ("C"), development schedules ("D"), engineering notes ("E"), memoranda ("M"), and reports ("R"). Forrester requested that each engineer submit biweekly reports but that proved beyond their tolerance, so he settled instead for bimonthly reports from his senior engineers. Those reports were consolidated and distributed to all staff members. He noted that reporting of "difficulties and delays" was particularly important because "some other member of the Laboratory may be able to assist in providing missing information or materials." He also went so far as to bring in an MIT English professor as a consultant in 1947.[27]

The ONR's increasing project oversight enhanced Forrester's documentary bias, now used to justify and preserve the project. One by-product of these investigations was the introduction of a new series of documents, the so-called limited distribution documents, or "L" series. Although these could and would be used later for classified materials, Forrester also used them as documentation of his ideas about the future of the project, as justifications of decisions or methods used by the project, and as defenses of the project against outside criticism. Robert Everett later noted that these reviews also served to focus their thinking and to ponder broader issues more deeply.[28]

Another response to MIT and ONR criticism was to accentuate the differences between Whirlwind and other computer projects of the same time period so as to justify Whirlwind's extraordinarily high cost. One of the important differences, as Forrester described in his L-5 memorandum of October 1948, was Whirlwind's focus on engineering and the importance of what Forrester called "systems engineering." Forrester's use of the term "systems engineering" may

MIT engineer Norman Taylor performs a component check of the Whirlwind five-digit multiplier, complete but without its panel doors.

have followed from a suggestion by Warren Weaver in his investigation of Whirlwind in early 1947: that Whirlwind's emphasis on engineering would likely be the emphasis for future computer development. In November 1947 Forrester convinced Columbia University mathematician Francis Murray that the scale of Whirlwind's engineering was what distinguished it from other projects like John von Neumann's Institute for Advanced Studies computer at Princeton.[29]

In response to MIT president Compton's request for more information in October 1948, Forrester described the concept of systems engineering. To Forrester, systems engineering required that designers always keep the final goal in mind, in this case an operational real-time computing system. Systems engineering involved "the knitting together of important and valuable systems from old and new components" to demonstrate "the useful application of research results." The nature of Whirlwind's original goals defined its end use: to perform general-purpose, rapid computations for a simulator, and to do so without error for long periods of time. That goal led Forrester to conclude that reliability was one of the primary problems to be overcome, and it prompted the significant differences separating his project from others.[30]

130

Forrester's systems approach led him to devote significant resources to improving individual component reliability. Because the computer used thousands of vacuum tubes and the system had to function for hours or days without failure, Forrester's group investigated tube manufacturers' methods, tested the tubes to determine their true lifetimes, and investigated the ways in which tubes failed. They then worked with the manufacturers to improve manufacturing processes so as to increase component reliability.

Although planning helped, Forrester's group also learned from their many reviewers, as in one case involving the so-called marginal checking technique. The idea came from a conversation between Forrester and Murray in the latter's investigation of Whirlwind in late 1947. Murray asked Forrester, "What are you going to do about all of these components when they gradually deteriorate...? You won't know you're approaching trouble until components begin to cause mistakes." Forrester hadn't thought of that but he replied that they could vary the input voltages on the tubes to check for faulty behavior by running standard calculations and making sure that the correct answer resulted. This marginal checking hardware made the computer more complicated than other machines of the time, but along with modifications to the manufacturing processes it led to improvements that extended vacuum tube life from around five hundred hours to five million hours.[31]

Another aspect of Whirlwind's systems approach was the primacy of Robert Everett's "block diagrams" group. Early in the design of the machine, Robert Everett and a small team of engineers developed what would later become known as the computer "architecture" through a top-level paper design and analysis. Everett's group apparently was unique in its documentation of these issues.[32]

Adopting the stored-program architecture then proposed for the EDVAC computer (the eventual long-term standard for general-purpose digital computers), Everett's team soon found that computation speeds would not be fast enough if digital information passed through the computer in a serial manner. To solve this problem, Everett adopted a sixteen-bit word, which included five instruction bits and eleven address bits. All sixteen bits were transferred in parallel, and the machine could compute numbers of thirty-two bits with two data transfers. Mathematicians from the ONR and elsewhere criticized this short word length because it would compromise calculation ability for highly complex mathematical computations. But for its intended application, periodic calculations of aircraft dynamics (and later positions) in a very short time period, the short word length sufficed and the parallel data transfer was essential. Parallel data transfer did lead to one significant design problem—the use of cathode-ray tubes for memory—and Forrester and his team never could make that method sufficiently reliable.[33]

Ironically, in the comprehensive review of computing systems done by Bell Laboratories' Nyquist and others for the RDB, the panel judged Whirlwind the worst design from a systems engineering standpoint. Given the commitment or at least the rhetoric of Forrester and his group to a systems approach, how do we now reconcile this contradiction? One option is to side with Nyquist's panel and judge Whirlwind's systems approach as mere rhetoric used to justify a poorly designed computer. Conversely, one could side with Forrester and believe that his group paid more attention to systems design, which led to the computer's unique characteristics. Taking one position or the other leaves the contradiction unresolved. A more promising tack is to conclude that the systems approach meant different things to Forrester than it did to Nyquist's panel, or that the systems approach meant the same thing to both but that Forrester's group optimized certain features of the machine that Nyquist's panel would not have chosen to emphasize.[34]

Nyquist's group stated that the military should develop computers for real-time control applications, yet somehow overlooked the fact that this *was* Whirlwind's purpose. Apparently in this case Forrester's argument that his machine could be used for many purposes was taken seriously by the reviewers. They criticized Whirlwind for not being designed for any particular application. Because

Norman H. Taylor, Robert Everett, and John A. O'Brien (shown left to right) inspect the Whirlwind I control matrix, 1951. Everett and his team of engineers developed parallel data transfer and what would later be known as the computer's "architecture."

by that time the project had dropped the development of the simulator to concentrate instead on the digital computer, it gave the impression of having no intended end application.

Whirlwind's troublesome cathode-ray-tube memories and clumsy mechanical structure came under Nyquist's criticism. For the structure the Navy had requested that Forrester contract out for the manufacturing, which led to the manufacturing contract with Sylvania. In this instance, Forrester's insistence on reliability led to a clumsy-looking mechanical design. He insisted on a two-dimensional layout to make every component available for replacement without unplugging anything. This yielded a machine that had all the elegance of an elephant. Forrester's application of the "systems approach" produced a monstrous mechanical design but one that had the key attribute Forrester desired—easy maintenance.[35]

The Whirlwind project stood out from other computer projects primarily because its cost was far greater. One of the primary reasons for this was Forrester's adherence to a strategy of concurrency, or parallelism. Having devised that strategy under wartime conditions when speed was of primary concern and costs unimportant, Forrester continued to operate in the same manner even after the war. Forrester insisted on having plenty of tools and components available so that these did not hold up the project. The project's criticality called for rapid development despite technical uncertainties. When the outcome of a particular approach seemed uncertain, Forrester pursued other possibilities in parallel. Another example of parallelism was Sylvania's manufacturing effort, which took place at the same time as the circuitry design. Although it speeded development it also cost more than would a leisurely paced project, just as Schriever's concurrency strategy would do for ICBMs a few years later.[36]

As Whirlwind became operational in 1949 and 1950 and became involved with air defense demonstrations for Valley's Air Defense Systems Engineering Committee, Forrester's group began to shift their efforts to the use of the computer. This involved communications with radar sites on one hand and with Air Force operators on the other. For the first time, Whirlwind engineers faced the problems of computer programming, and soon they found that to program the computer they would have to understand not only the hardware characteristics of radar systems, telephone lines, and cathode-ray-tube displays, but also the tactics of air defense and the characteristics of human operators.

Learning to Develop a System

By the summer of 1950 Whirlwind was operating and, with Valley's Air Force funding through ADSEC, Forrester's group began taking the first steps from computer hardware development to air defense system development. Computer hardware development never stopped but Forrester's group expanded its activities beyond hardware to software development and large-scale systems integration to understand the requirements for the inputs and outputs of the computer and to program the computer for air defense applications. That required the group members to master new skills and processes beyond the electrical and mechanical engineering necessary to design the hardware.

Valley and ADSEC neatly summarized many of the problems they would face in equipping the computer for air defense in a long memorandum to the commander of the Air Force Cambridge Research Laboratory in October 1950. They tried to pull together the various analyses and studies done by ADSEC into a coherent framework that could be used to introduce the subject. The paper's writers decided to use the concept of "system" to illuminate the topic. To them, the air defense system was "an organism," a "structure composed of distinct parts so constituted that the functioning of the parts and their relation to one another is governed by the relation to the whole." They noted that organisms came in several kinds: animate organisms like animals, animate organisms with inanimate devices like the air defense system, and inanimate organisms such as vending machines. Most important, all organisms contained several functional components in common: "sensory components, communication facilities, data analysing devices, centers of judgement, directors of action, and effectors, or executing agencies."[37]

Illuminating their paper with military examples, such as Caesar's Army as a representative organism, Valley and ADSEC then organized their description of the air defense system in terms of the functional components listed above. The use of the systems metaphor allowed Valley to describe each element of air defense in terms of communication and information. For example, for local radar sites of the period he claimed that the "orderly" arrangement of information from the radar was lost when sent to cathode-ray tubes for operators to interpret. Even worse, the interpreted data then were sent through voice telephone lines, which caused the loss of much of the information originally available from the radar and left the data potentially misinterpreted or incomplete. A better approach was to send the original data directly to a centralized location for interpretation and analysis, thereby preserving all of the original information. The centralized approach prevailed and the Whirlwind team went to work to make the computer at its core sufficiently capable and reliable in handling the input.[38]

To develop this "organism," MIT administrators originally organized Lincoln Laboratory into five divisions: 1–administration, 2–aircraft control and warning, 3–communications and components, 4–special projects, 5–reconnaissance. Because of its prior history, existing organization, and stubborn independence, Forrester's group remained for a time part of the Servomechanisms Laboratory of MIT. However, the group soon transferred to Lincoln and became Division 6. By the summer of 1952 Division 4 focused on weapons, Division 5 on "special systems," and a new Division 7 on "engineering design and technical services."[39]

MIT's Jay W. Forrester, the inventor of the magnetic core memory, is shown holding a 64x64-digit memory plane. The plane was used in the Whirlwind/SAGE computer.

Forrester's Division 6 faced two major problems. First, it was becoming clear by the summer of 1951 that the cathode-ray-tube memory simply was not going to be reliable enough to provide the round-the-clock service required for air defense. An alternative had to be found. Second, the Whirlwind machine had been Forrester's "pre-prototype" and it was insufficient to handle the complete air defense problem. His team would have to design another machine with higher performance and more memory.

Finding a solution to the memory problem was the first order of business. Indeed, it was pivotal to a digital computer-based air defense system. Unhappy with cathode-ray-tube memory reliability, Forrester began searching for alternatives in the late 1940s. Although others had considered them impractical, Forrester vigorously pursued the use of magnetic materials and much to others' surprise, he made them work. He developed the idea of using small doughnut-shaped ferrite (magnetic) materials known as magnetic cores that could be wired into a three-dimensional matrix. By reversing the polarity of the magnetic materials through electrical signals, each core represented one "bit" of information, either a zero or a one. After substantial research and experimentation starting in 1950 and a search for suitable manufacturers, Forrester successfully demonstrated a sixteen-by-sixteen array of ferrite cores in May 1952. (Independent of Forrester, IBM had come up with a similar idea and demonstrated it that same month.) This innovation overcame the key technical hurdle and significantly enhanced the reputations of Forrester and his team. As George Valley put it later,

"After I'd seen the first satisfactory cores, my attitude towards the Whirlwind people changed. I began to take them seriously and to regard them as worthy of respect."[40]

The second major issue was designing and manufacturing a more powerful computer to handle the full air defense load. In 1952 Forrester's staff established the Whirlwind II group to design the new machine. To create an appropriately capable computer they had to estimate the number of computations that would be required, and that prompted considerable discussion and negotiation with Air Defense Command and Air Research and Development Command. Major criteria included the allowable downtime for the machine, the number of tracks to be processed, and the number of interceptors that could be deployed simultaneously. Based on this information, the Lincoln group determined the performance and reliability necessary for the machine. Because many reproductions of the computer were to be installed in locations (sectors) around the United States and Canada, Forrester's team recognized the need to engage a computer manufacturer to produce the machines in quantity. That need led to an evaluation of vendors in Forrester's typically thorough fashion—creating a matrix of desired characteristics beforehand and then evaluating each vendor against these preestablished criteria. The team selected IBM in October 1952.[41]

Originally called Whirlwind II, the Lincoln–IBM team eventually renamed the prototype the XD-1 and called the production machine by its Air Force designation, the AN/FSQ-7. Originally called the Transition System, in 1954 the overall air defense system was renamed the Semi-Automatic Ground Environment, or SAGE. From the early beginnings of Whirlwind to the SAGE system, Forrester's team, augmented by IBM's substantial capabilities, created a series of organizational innovations to organize computer development and its associated programming.[42]

With the considerable experience they had gained in designing Whirlwind and its application to air defense, Forrester's Division 6 engineers had no intention of letting IBM dominate the design of the AN/FSQ-7. In fact, they planned to design the computer and then hand over the design to IBM engineers, who might make minor modifications for production purposes. By contrast, IBM's engineers were fresh from the design of the highly successful 701 computer and considered themselves the preeminent experts in computer design. They expected to design and deliver the AN/FSQ-7 as they had always done in the past. In practice, two headstrong and highly competent organizations would butt heads and have to discover how to work together and learn from each other. Their first meetings were "loud and rancorous" but eventually the teams came to respect each other's abilities.[43]

Knowing that the design would be handed over to a contractor, Lincoln engineers collected design information at a central location and concluded that they would have to exercise some form of centralized control over the design. Forrester and Everett signed off on all block diagrams of the design and resolved all

Control consoles of the XD–1 computer for SAGE experimentation. Originally, the XD–1 was known as Whirlwind II, but it was renamed by the Lincoln–IBM team.

conflicts within Division 6. To facilitate and control communication between Lincoln and IBM they also established a single group to track all correspondence and communication between the two facilities. The Lincoln group organized itself according to major subsystems, such as arithmetic element, memory, drums, and so on, and IBM did the same to facilitate communication.[44]

Formal mechanisms could document design and communication but could not ensure cooperation between Lincoln and IBM. To meet the Air Force's ambitious goals the two organizations needed to go beyond formal communications. To promote expanded cooperation the two organizations began a series of meetings to get acquainted with each other and to help IBM learn more about the requirements of the project. Starting with meetings halfway between Bedford, Massachusetts, and Poughkeepsie, New York, where IBM would manufacture the new computers, the two groups met in Hartford, Connecticut, in January, April, May, and June 1953. During the first meeting both organizations agreed to break into subsystem committees. The committees reported on their status and anticipated roles during the second meeting. The next three meetings continued the technical interchanges and consensus building on design issues such as packaging, cooling, and interconnection of major components. Following the Hartford meetings were seven more meetings between IBM and Lincoln engineers, known

Rear view of the internal wiring of the AN/FSQ-7 manual inputs patch panel.

as Project Grind, to grind out the technical details of the major subsystems. These took place in June and July of 1953 and resulted in consensus on many but not all of the necessary technical features.[45]

The Hartford and Grind meetings resulted in many important decisions by the committees in charge of each subsystem, but there remained the task of coordinating each of those decisions with other subsystems to ensure their mutual compatibility. The problem was that no group had overall responsibility for approving design decisions across the entire air defense system. The Air Force had formal authority to make final overall design decisions but the real decision makers—IBM and Lincoln—did not. MIT's management interpreted Lincoln's charter to be "research," whereas IBM's task was "production." What was missing was a group that acted as the "architect" of the entire system.

John Jacobs in the Whirlwind II group (Group 62) recognized the problem, and began informally to gather information and justifications for each circuit element. He then circulated that information to the different committees to get their official sign-off that the design was satisfactory. After that, he and a small group of engineers wrote a brief summary of the design decisions and final design and circulated them through Lincoln Divisions 2 and 6, the Air Force's Lincoln project office, and Air Force representatives at AFCRC and ADC. The briefs drew conclusions "that tended to be the least unacceptable to any of the affected parties" in order to achieve a consensus. Jacobs then circumspectly "solicited the signatures of those who were in a position to block the action." By the fall of 1953 Forrester and Everett created the systems office under Jacobs to continue this coordination and consensus function.[46]

Jacobs had to pay as much attention to the politics of the situation as to the technical issues. The problem was that Lincoln did not have the authority to perform these duties and it would not do to call attention to this fact, either from MIT or from the Air Force and its potential critics. As he later noted,

> we could not put out a directive because that would force the organization to determine who was in charge. . . . Although the Air Force had all the authority it needed to designate someone in its own organization to be responsible for the design, this responsibility was precluded by the fact that the Air Force did not have the technical capability to effectively monitor the design as it progressed, much less initiate it.[47]

Jacobs' diplomatic approach to the problem of coordinating technical activities allowed all parties concerned to make progress while avoiding the sticky problem of authority. In essence, Lincoln's control of the project could only be sustained as long as development and communication remained informal. Legally, the Air Force had to be in charge because it distributed the taxpayers' money to develop SAGE. Air Force leaders recognized the problem and tacitly encouraged all parties to acquiesce in the unusual arrangement.

IBM quickly mirrored Jacobs' organizational innovation by creating their own Engineering Design Office to work with Lincoln's systems office. Robert Crago, the engineer placed in charge of the Engineering Design Office, worried that the cumbersome consensus procedures slowed the design effort excessively. As Jacobs later put it, Crago spent much of his time "loosening the grip of Lincoln on IBM operations." He did this by insisting that "the less important decisions should be the prerogative of the IBM people in charge of the subsystems." Crago's success at speeding IBM's design led eventually to his promotion to head of the AN/FSQ-7 project at IBM.[48]

The need for documentation and approval of documents for the entire project led to the institutionalization of Jacobs' coordination. Jacobs' group developed a standardized form, known as the Technical Information Release, or TIR. It ultimately contained a package of information, including design specifications, background data, analyses, and recommendations and their supporting materials. This method of "design control" required three steps tracked through the systems office: preliminary status, concurrence between the two engineering organizations, and final authorization and release for use in construction.[49]

Jacobs searched for a word or phrase that would describe the process without stirring up trouble. Eventually he adopted "concurrence" as appropriate to the situation. As he explained his choice,

> The words "direct" and "control" were too strong. "Oversee" and "coordinate" were too weak. But concurrence had the right feel—it gave everyone who was named as necessary to support the technical information release the implied power to veto what was being included in the TIR.[50]

While Jacobs struggled with the problem of coordinating the diverse activities of AN/FSQ-7 hardware design, engineers slowly began to recognize that an equally difficult problem involved the instructions needed to make the computer display and control radar data. These instructions eventually would be called "software." To make Whirlwind an effective central element in Valley's air defense "organism," Forrester's engineers would have to understand the system's components and the tactics and methods of air surveillance and combat. Based on those data, Lincoln engineers then had to provide the relevant information and options to Air Force operators. The bulk of the information would be encoded as computer instructions. Because AN/FSQ-7 was one of the first stored-program computers in operation in 1950, few precedents existed for this novel task, soon to be called "computer programming."

Individual engineers working on specific elements of the computer, the air defense problem, or peripheral devices wrote the instructions—that is, the programs—they needed to test their components or demonstrate the capabilities they were charged to create. Each engineer learned to program the machine in machine language (because no computer programs existed to translate a higher-level language into machine language), and each one squeezed programming time in between upgrades to the computer, maintenance, and integration of other hardware with Whirlwind.[51]

These "proto-programmers" fought for time on the computer because it still used troublesome cathode-ray-tube memory. Forrester's group worked frantically to resolve memory reliability problems. Between the spring of 1950 and 1951, tube reliability had improved from less than an hour to several hours. With this gain, Forrester considered the Whirlwind "a reliable operating system." The computer operated on a schedule of 35 hours of applications programming a week and it achieved that goal 90 percent of the time. It ran as long as seven hours without failure, a new record for the machine.[52]

With computer time scarce and substantial amounts of programming to accomplish, C. Robert Wieser, in charge of the design and testing of the prototype "Cape Cod System," decided to formalize the ad hoc organization that characterized early programming. The Cape Cod System would integrate Whirlwind with radar and communications gear to test and demonstrate the capabilities of

Direction center of the Cape Cod System, which integrated Whirlwind with radar and communications gear to demonstrate the capabilities of the three technologies by detecting real aircraft, portraying their tracks on cathode-ray-tube screens, and performing interceptions.

integrated computer–radar–communications technologies by detecting real aircraft, portraying their tracks on cathode-ray-tube screens, and performing interceptions. In February 1951 Wieser grouped his engineers and mathematicians as analysts, programmers, and operators. The analysts focused on what the system was supposed to do and wrote down detailed specifications defining those functions, using standard English supplemented with mathematical equations. The programmers encoded the specifications into computer "orders"—that is, instructions or computer code. Finally, the operators concentrated on operating the entire integrated system.[53]

By December 1951 computer hardware reliability had improved significantly. Through 1950 and most of 1951, frequent hardware problems led programmers first to blame the hardware for difficulties in operating their programs—blame that bore some justification. John Gilmore of the Applications Group took a different tack, however, and exhorted programmers to look instead to flaws in their own logic. He noted that almost all problems were in the programmer's code and not the hardware. He said that the programmers' mistrust of the computer was unjustified.

Instead of having each programmer run the machine on his or her own, Gilmore put two people in charge of running the computer. Programmers had to submit their programs to the operators with sufficient documentation so that the operators could set up the correct inputs and outputs and know if the program had gotten into an infinite loop or some other malfunction. This information would be documented on new program performance request forms. Because many applications programs used similar mathematical and utility algorithms, Gilmore and the programmers began to create libraries of standard routines using coding conventions recently developed for the University of Cambridge EDSAC computer in England. Gilmore encouraged programmers to contribute to the growing library of reusable subroutines.[54]

Programmers learned that they had to manage carefully the writing and testing of Lincoln's increasingly large and sophisticated software programs. By the summer and fall of 1953, programmers learned to split their programs "into very small sections which were checked individually" using data from tapes and stored constants to stimulate the algorithms. As sections (what we now would call subroutines) passed their individual tests, "they were joined into larger sections and restored until a single program resulted." For example, programmers divided the "Non-Track While Scan" program into approximately fifty smaller programs that they tested individually. After successful completion of these tests they combined the fifty small routines into four larger programs and tested those. Finally, they combined those four into a single routine that they tested as a unit. The tests themselves included so-called static tests that used only simulated inputs and dynamic tests that included use of the actual input–output equipment in the air defense direction center.[55]

To support the growing number of programmers and programs, MIT's programmers created a substantial number of "utility" programs to assist with

The Flexowriter, a SAGE input– output device.

The heart of the AN/FSQ-7 air defense computer was the magnetic core memory, which contained more than 500,000 bits of high-speed storage with an access time of six-millionths of a second to any group of 32 bits. The memory held instructions that the computer would follow during its normal operations and the data being correlated and processed at that time. Shown here from left to right are the computer's memory array frame, the memory frame, and the power module.

programming tasks. This effort was intended in part to make coding and testing more reliable and in part to speed up those tasks, given the severe limitations on computer time available for each programmer. The utility programs included a "manual-intervention program" to permit the "inspection or change of the contents of the individual registers, both in internal storage and on the drum."* They also included programs for "printing out selected blocks of orders and data for analysis purposes" and for starting and stopping programs.[56]

As the Whirlwind project progressed from computer hardware design to the application of computers to air defense, Forrester and his organization had to develop new methods to organize and coordinate their activities. Two features of this change were particularly important: the selection of and interaction with IBM as the production contractor for the new AN/FSQ-7 computer, and the

---

*The drum was a form of long-term storage similar in function to "hard disk" storage on current computers.

A close-up view of the
AN/FSQ-7 magnetic
core memory array
frame, the heart
of the computer.

increasing significance of computer programming. Relations with IBM meant that Forrester's group had to share decision making with another group and to formalize communications with an external organization. In the peculiar circumstances whereby the Air Force tacitly delegated authority to Lincoln Laboratory, Lincoln's engineers had to tread softly.

Computer programming was an entirely new function that required developing new tools, communication methods, and organization. Because the computer program for the new system encapsulated the Air Force's strategies for air defense as well as information about all of the devices and operators to which the computer connected, the computer programmers acted as integrators of all other components and their associated organizations. In these formative years, managers and programmers noticed how the problems they faced grew in size and complexity. Their immediate response was to place more people on the problem, to create tools to help the programmers, and to formalize their organization and methods. Lincoln's understanding, organization, and communications grew increasingly sophisticated, and that would serve them well as the scope of their tasks and interactions with the outside world increased dramatically.

Semi-Automatic Air Defense

As MIT engineers developed new methods to coordinate the application of computers to air defense, the Air Force created and modified organizations to coordinate these efforts into the Air Force's R&D programs and operations. ARDC Headquarters assigned to AFCRC the responsibility for engineering the Transition System, with Lincoln Laboratory acting "as a consultant to the Air Force through AFCRC." Thus the Air Force officially considered Lincoln's work, whether in research, development, or testing of all or parts of the system, to be "advice." AFCRC in turn assigned a team of fifteen people, including John V. Harrington as the project engineer, Lt. Col. Benjamin Blasingame as the assistant project engineer, other technical experts, contractors, and clerical help.[57] To handle procurement of Lincoln Laboratory developments, ARDC Headquarters requested that AMC assign and colocate personnel in the Cambridge area to work with AFCRC.[58]

Whereas, on the one hand the Air Force arranged to work with Lincoln Laboratory, on the other hand it also hedged its technological bets. Many officers counted on the Air Force's bombers to deter attack, some questioned whether MIT's automated scheme would work at all, and still others thought that automation was the wave of the future but was unrealistic for the present. In that last group was ADC Commander Gen. Benjamin Chidlaw. In October 1952 he recommended that the MIT project be oriented toward future defense against ballistic missiles, and that in the near term the Air Force fund the University of Michigan's less ambitious scheme. To promote the idea, he planned a test of the Michigan system in late 1952.[59]

In the air surveillance room at Stewart Direction Center, Air Force personnel constantly monitored radar data from many remote sites. The information was transmitted over telephone lines to the network of direction centers.

The USAF and the Culture of Innovation

Michigan's Air Defense Integrated System (ADIS) project differed from MIT's approach in a number of ways. Whereas MIT's trajectory and interception calculations occurred in a few centralized SAGE computers taking data from many radar sites, ADIS performed the calculations locally at the radar sites with analog computers and individual operators who sent this information the old-fashioned way to a direction center. The system depended on tried-and-true technologies. Michigan researchers planned to automate the transfer of local data from place to place to meet the objections of those who favored the centralized MIT approach.[60]

ADIS supporters knew how to woo the Air Force's midlevel officers. MIT depended on its World War II Radiation Laboratory experience, prestige, and technological prowess to convince the Air Force's leaders of its credibility. Its pitch essentially was "we've done this before, we understand what you need, we have the smartest people, and you should trust us." The Air Force's leaders, who had been through World War II, knew personally and trusted MIT's Radiation Laboratory veterans. By contrast, ADIS personnel had no such advantage so they worked with ADC's midlevel officers to determine the number of operators that would be needed, "the location of the general's command post, and the human engineering of displays and devices for entering data into the system." By MIT's standards they paid little attention to developing new hardware and they studied their system through simulations on an analog simulator. Thus many operational officers understood ADIS operations but not its underlying technologies. They did not have a corresponding understanding of Lincoln's system because Lincoln personnel focused on developing the radar and computing equipment instead of how officers would work with it.[61]

Some of Lincoln's supportive Air Force officers went to talk about the situation with George Valley, and some of IBM's sales personnel met with Valley and Forrester. The IBM representatives, wise to the ways of commerce, told them, "in our business we've discovered that it is necessary to give the customer a little of what he thinks he wants, in order to maintain oneself in a position to give him what he really needs." After those meetings Valley pressed his Lincoln associates to work out some of the operational details the Air Force's working officers wanted. Lincoln leaders were frustrated that the Air Force's officers could not seem to distinguish between Michigan's preprogrammed demonstrations (a pure simulation) and Lincoln's demonstrations using real aircraft, radar, and a digital computer. Technology supporters like Ivan Getting were equally appalled by the lack of progress at the Rome Air Development Center, which managed the Michigan effort. He tartly remarked that they had "just found out where the bathroom was, much less how to run a new development."[62]

By late 1952 the MIT administration, concerned with the Lincoln Laboratory's finances and with the threat that Michigan posed to the project, decided to press the Air Force to decide between the two systems. In January 1953 MIT president James Killian wrote to Secretary of the Air Force Thomas J. Finletter

146

to request another system review led by Dr. Mervin J. Kelly, the president of American Telephone and Telegraph Company. More significantly, Killian wrote that MIT normally did not take on these kinds of projects. He threatened that

> we stand ready to withdraw since the Project involves many hazards for the Institute, particularly financial hazards, and since it is not the kind of Project the Institute normally would wish to undertake, we feel it important that there be no question whatsoever with regard to our serving as contractor. From the standpoint of the Institute's interest, it must be said that it would be better for us not to be the contractor.[63]

Killian's letter was not merely an idle threat because a financial crisis in late 1951 made MIT's administrators and trustees uneasy. In late 1951 the Navy reduced some of its contributions for the Research and Development Board, which funded a portion of Lincoln's activities. This caused a funding problem right at the start and made MIT's leaders wary of taking on further financial commitments without a corresponding commitment from the Air Force.[64]

Finletter initially responded by defending Air Defense Command and their support of Michigan, stating that "there are other groups who have ideas on Air Defense and equipments for Air Defense that will probably be available before the Lincoln Project can provide any such. We feel it is our duty to support such efforts but assure that they will not detract from the Lincoln program."[65]

ARDC Commander Lt. Gen. Earle Partridge wrote in further explanation to Killian, "On the basis of this present meager knowledge, the Air Force is unwilling to commit itself to a large scale production program on either system to the complete exclusion of the other." Referring to Michigan's ADIS, he stated that AMC was establishing a contract "to procure limited production quantities of all components through a single prime contract." He wanted progress to be made on both the MIT and Michigan systems so that in nine months the Air Force could make a decision "on either or both."[66]

The government's financial ineptitude defeated the Air Force's deliberate strategy by handing MIT more arguments to force a quick decision. Although the Air Force supplied millions of dollars, the government's way of doing business was often fickle, particularly at the July 1 beginning of each fiscal year. By February 1953 MIT and the Air Force were concerned that funding for the project was going to run out at the end of June, although the contract was valid through December. The deficit in funding that MIT would be forced to carry would have to be covered by MIT endowment funds. Alarmed at the prospect that the Air Force would live up to MIT's worst expectations, AFCRC's commander warned ARDC Vice Commander Donald Putt that "in the past when M.I.T. has had to resort to this, repercussions have resulted." MIT president Killian could pressure the Air Force to commit wholeheartedly to their project, or MIT would back out.[67]

Although the details are not clear, this financial crisis undoubtedly was one of the factors that led the Air Force to "initiate a unilateral approach" to air defense by working solely with Lincoln Laboratory. Citing World War II experience that optimal air defenses at their best shot down only 6 to 8 percent of a bomber force in each attack, Air Force leaders did not believe that air defense would ever provide a foolproof shield against nuclear attack. In light of that experience, they put their trust in deterrence through SAC's bombers. They refused to make air defense the centerpiece of their technology efforts and would not fund two such systems. Forced into an early decision, senior leaders ultimately believed MIT's Radiation Laboratory veterans over Michigan's newcomers. They canceled ADIS in May 1953 and funded Lincoln with the $300 million allocated for Michigan. ARDC's Putt told the AFCRC commander that AFCRC would "carry out system engineering responsibilities" for ARDC, including integration of other weapons such as the BOMARC missile.[68]

Integration of interceptor aircraft and missiles required that Lincoln and AFCRC communicate with the organizations that built and managed these weapons. Rome Air Development Center had been working since 1951 with the Army's Integrated Anti-Aircraft Fire Direction System through the Army Signal Corps Engineering Laboratories in Belmar, New Jersey. With the decision to abandon ADIS, AFCRC replaced Rome as the coordinating center in May 1953. Lincoln had already been working with the Army so the transition from Rome to AFCRC was not difficult.[69] During the summer of 1953, AFCRC organized a series of meetings between their officers, Lincoln personnel, and the other organizations that would have to be involved with air defense, including Boeing with its BOMARC missile, and Hughes Aircraft and its F–102 interceptor programs run through Wright Air Development Center.[70]

AFCRC also organized a critical meeting with Air Defense Command to work out technical details and determine costs for installing the Transition System with an ADC wing. AFCRC and Lincoln prepared meticulously for this critical encounter, going so far as to hire retired ADC Commander Gen. Gordon Saville as a consultant to help them prepare for the meeting.[71] Despite the concerns of AFCRC and Lincoln personnel, ADC personnel bore no apparent hard feelings about losing the battle over ADIS and they quickly oriented themselves to work on the Lincoln Transition System, which was now renamed the Semi-Automatic Ground Environment, or SAGE.[72] ADC established the 4620th Air Wing at Hanscom AFB in Lexington, Massachusetts, to work with the Lincoln staff. The wing was commanded by Col. Joseph Lee. John Jacobs, head of Lincoln's systems office, soon met with Lee and negotiated an arrangement whereby Lincoln gave him "veto power" but not "directive authority" over Lincoln's activities. Thus Lincoln's technical people would coordinate and obtain consensus with ADC just as they did with IBM and AFCRC.[73]

The SAGE system of continental air defense was announced at a press conference in January 1956. Hosting the conference at Lincoln Laboratory were (left to right) Edward L. Cochrane, MIT vice president for industrial and governmental relations; George E. Valley, associate director of Lincoln Laboratory; and Maj. Gen. Raymond C. Maude and Col. Dorr E. Newton, Jr., respectively the commander and vice commander of the Air Force Cambridge Research Center.

While Lincoln made organizational arrangements to continue technical decision making, AFCRC, a branch of Air Research and Development Command, negotiated with Air Materiel Command regarding the financial and contractual aspects of the program. This led to the cumbersome and somewhat touchy arrangement whereby the Rome Air Depot of Air Materiel Command negotiated with IBM on behalf of ARDC, even though the real work and negotiations took place between Lincoln and IBM.[74] Facilities for the AN/FSQ-7 required further coordination, this time with AMC's Industrial Facilities Office at Wright-Patterson AFB. In this case, Lincoln Laboratory representatives requested that IBM submit facilities cost information to AFCRC, who then passed them to AMC.[75]

The initial concern of ADC leaders was to organize and allocate the SAGE system's radar and computers across the continent. Given the available and projected radar and computing capabilities, discussions among Lincoln Laboratory, AFCRC, and ADC personnel ultimately yielded nine sectors and forty-two subsectors. Each sector would house a "combat center" and each subsector would have a "direction center." To oversee the construction of the facilities and the production and installation of the facilities, communications gear and electronic equipment, Air Force leaders selected Bell Telephone Laboratories (BTL) and Western Electric (WE),† to be managed by a new AMC office located in New York City and known as the Air Defense Engineering Services (ADES). BTL and WE had performed well on the Continental Air Defense System project to install and link radar facilities with Air Force direction centers since 1950, and that past performance led to their selection to perform a similar task for SAGE. Although Lincoln would lead the R&D effort to develop, integrate, and test the new system, IBM, BTL, and WE would produce and install that equipment across the continental United States.[76]

Fresh from projects in which they had full control of air defense development, the AT&T companies, BTL and WE, initially believed or hoped that they would have the same level of control over SAGE. Perhaps not understanding (or not wanting to understand) the unique, informal relationships that existed, BTL soon tried to gain control over Lincoln. Preferring a more decentralized approach reminiscent of Michigan's ADIS, some ARDC officers who wanted to rein in Lincoln's independence supported BTL's attempt. Both ARDC and BTL hoped to push Lincoln toward research on "radical concepts for the future" and hence gain control of the current development of SAGE. Unfortunately for Bell and ARDC, Bell's effort failed when Lincoln's experts exposed Bell engineers' ignorance of the details of the SAGE program and the testing of the system.[77]

At a meeting in July 1954, Western Electric finally agreed that Lincoln held design responsibility for the system, including "authority to dictate a) the design, b) system performance, c) test requirements, d) design changes, etc." If Lincoln was to hold this authority, then Western Electric's spokesman also placed the responsibility on Lincoln. He made sure that Lincoln would provide the requirements, tests, and test methods to be used and would ensure that the Air Force agreed to them. Western Electric, he stated, "will give assistance as required by Lincoln" to prepare test specifications and would carry through the test program after Lincoln and his company worked through the initial installations. Lincoln agreed to define and control a "well-defined procedure for handling design changes." Where costs and delivery schedules were unaffected Lincoln would have full authority, but the Air Force Joint Office would have to approve any circumstances in which costs, schedules, or operational requirements were affected. Western Electric's spokesman acquiesced to Lincoln's primacy, but he warned that "while Lincoln could delegate design and test specification duties to IBM, BTL or Western Electric or others, Western Electric would hold Lincoln responsible."[78]

With the peace agreement between Bell and Lincoln, the SAGE program got fully under way. Throughout 1954, Lincoln, AFCRC, and the 4620th Air

---

†Western Electric was the manufacturing arm of the Bell System, or American Telephone and Telegraph; Bell Telephone Laboratories was AT&T's major R&D center.

Wing ran experiments on the Cape Cod System to determine how best to operate it. They determined such things as the proper lighting, operator commands, and user interfaces to the computer system, along with technical issues related to the communication, reduction, and display of radar data. In parallel with this, Lincoln created an organization to prepare for the "Experimental SAGE Sector," which would host the first prototype computer known as the XD-1. Forrester's Division 6 also created a production office that paralleled Jacobs' systems office to coordinate between Lincoln and all of the new organizations involved with the design, production, and installation of the continent-wide air defense network. Those organizations now included AFCRC, ADC, Bell Laboratories, Western Electric, and the ADES office.[79]

Among the concerns of the ADES office, and particularly its representatives from Air Defense Command, was the need for field trials and for the training of Air Force personnel to operate the new SAGE system. That concern led to a visit by RAND Corporation personnel to Lincoln in late December 1954. RAND Associate Director J. R. Goldstein and Melvin O. Kappler led a small contingent of RAND personnel to describe their work on simulation of air defense operations and its use in training Air Force personnel. Lincoln's interest was to see if RAND could supply inputs or expertise to help test SAGE. Frank Heart, who hosted the visit for Lincoln, concluded that for the moment it did not appear like it was worth the effort to have the RAND personnel help with Lincoln's simulation. He noted, however, that it might be worthwhile later for them to develop or assist with training exercises for SAGE. In fact, RAND and its descendent, System Development Corporation, soon would become the major contributor to computer programming for SAGE and its successors.[80]

RAND Enters the Scene

RAND's involvement with SAGE stemmed ultimately from a psychologist's search for research that might be of interest to the Air Force. In August 1950 psychologist and RAND consultant John L. Kennedy suggested a research project to study interactions among individuals in a realistic but controlled setting. Kennedy, who had been a consultant to the Office of Scientific Research and Development's Applied Psychology Panel during World War II, became interested in the experiments of Alex Bavelas at MIT. Bavelas had performed experiments to study social "group structure as it affects the output of an organization."[81] He was a student of renowned social psychologist Kurt Lewin and completed his dissertation in 1948.

The experiment that so interested Kennedy was a study of communications among five individuals to investigate how different communications "connectivity" affected task performance and learning. Typically Bavelas worked with five subjects, who in the experimental setup could not see or hear each other but could communicate with each other through written messages and connected with each other in different "networks." For example, one communications network would be a linear chain along which each individual could communicate only with two others, one on each side. In another case, Bavelas would make one person the "central communication node" connected to each of the others, who were connected only to the central person. Bavelas' main accomplishment was to turn a complex social situation into a controllable, quantifiable psychology experiment. This innovative experiment drew admiration from other social psychologists and organizational theorists.[82]

Kennedy proposed to expand on Bavelas' five-person experiment. He noted that the Air Force's air defense system was a human–machine network, albeit much larger than Bavelas' tiny group. While engineers could predict the performance of the machine portion of the system, psychologists could not do the same for the human portion. Furthermore, Kennedy stated that "we have a great deal of difficulty as psychologists in extrapolating our knowledge of individual behavior to the group situation." He thought the time ripe "to begin to generate the kind of hard-number information about people behaving in groups and organized in networks." Kennedy proposed to perform experiments "with larger and more realistic groups of the Bavelas type, utilizing a more realistic task so far as the air defense system is concerned." Basing their interest in this experiment in part on Melvin Kappler's interest in air defense equipment, Kennedy and his colleagues decided to simulate an air defense direction center.[83]

To facilitate the experiment, known initially as "Project Simulator," Kennedy's group founded the Systems Research Laboratory (SRL) in May 1951. The laboratory was charged with studying "particular kinds of models—models made of metal, flesh and blood." Using an information processing center model, Kennedy's group would study the information flow between individuals and machines in their direction center mock-up. The object of the experiments was to determine how human operators learned to accomplish the air defense task. Key to the analysis was the "organism," or "system concept." To Kennedy's group, the basic problem was that in a complex situation that included several humans and machines there was no way cleanly to separate and isolate learning to individuals. Instead, they would stimulate the group and analyze the interactions as a whole.[84]

The RAND psychologists realized that their experiment had direct bearing on the managerial function. Their goal was the same as management: to achieve the optimum allocation of tasks, resources, and processes among humans and machines. Through laboratory experiments, they would study the interactions among individuals and machines to help managers determine the best allocation of resources and how best to motivate the individuals in the organization.[85]

The laboratory itself consisted of two large rooms with a separate viewing deck above and along the full length of one wall. Experimenters in the observation deck observed activities in the laboratory, and the telephone system and microphones picked up all of the conversations. The experimenters then could analyze and classify all conversations and activities. Using standard "stimulus–response" behaviorist learning theory, they proposed to stimulate the group using simulated aircraft tracks as they would appear on a radar scope. The group's expertise with computers developed because Allen Newell, the group's mathematician, programmed RAND's IBM 604 computers to calculate the aircraft tracks and superimpose the positions onto a series of paper sheets that would flip progressively through a mechanical viewing device. Later, RAND analysts programmed computers to classify and present the resulting data as well.[86]

Working with officers from Air Defense Command, Kennedy's group simulated the operation of McChord Air Force Base near Tacoma, Washington, using a crew of twenty-eight students recruited from colleges near Santa Monica. The SRL crew ran the experiment, known as "Casey," from February 4 to June 8, 1952. As the experiment proceeded the researchers were astounded to find substantial increases in the performance of the crew. To assess the learning capability of the crew, the experimenters systematically had increased the complexity of the tasks, and they found that the students organized themselves to account for and master the increased workload.[87]

Even before the end of the Casey experiment, the group recognized the potential utility of this finding for the training of Air Force crews. In late May 1952 RAND director Frank Collbohm wrote to Maj. Gen. Frederic Smith, the vice commander of ADC, regarding the potential use of the experimental environment for crew training. As Collbohm noted,

> The attainment of this kind of training program has been hampered by the difficulty and expense of providing a realistic and controllable radar input to the ADDC [Air Defense Direction Center]. However, the Systems Research Laboratory at RAND has found a technique for generating and preparing a sufficiently characteristic input in a reliable and economical fashion.[88]

Collbohm described the mechanical device for feeding paper at the same rate as a radar scope sweep and the capability of the SRL to generate a variety of flight tracks to place on the paper. The key was the use of "high-speed computers" to calculate the positions of aircraft, and "print a facsimile of the scope face on a section of a continuous sheet of IBM paper." In short, RAND's newfound ability to generate simulated aircraft trajectories through computer programming had lowered the cost of realistic simulation dramatically.[89]

The SRL staff briefed officers at ADC Headquarters in Colorado Springs in August 1952. They generated sufficient interest for ADC to supply a crew of officers and enlisted men for a second experiment, with Air Force personnel as subjects. This second experiment, known as "Cowboy," took place in January and February 1953 and yielded similar results. The Air Force team learned to cope with situations that would have been impossible without such training.

Smith and thirteen other officers from ADC Headquarters observed portions of the experiment and came away impressed.[90]

Briefings in March 1953 led to a joint RAND–Air Force study group charged with determining how to transfer the results of the experiment to a full-scale program to train all Air Defense Direction Center crews. By May the group recommended that RAND be involved with the effort because of the expertise of the Systems Research Laboratory, and by August 1953 the Air Force signed a $1.2 million contract to develop a System Training Program. Before full-scale commitment, Air Force Headquarters required a field test in one Air Force division. This led to a substantial increase in the SRL staff and in other RAND divisions to support the crash program. To indoctrinate the new staff, another test run called "Cobra" took place in February 1954 using another military crew. A fourth run, known as "Cogwheel," ran for two weeks in June 1954 to train Air Force officers in the principles and practice of system training.[91]

To prepare for these experiments and for the field trials that began in August 1954, the SRL programmed RAND's new IBM 701 computer to calculate trajectories, which operators viewed on a sophisticated scope simulator with the same look and feel as the real scopes. The Air Force planned to install the System Training Program at 152 sites, leading to the separation of the "System Training Project" from the System Research Laboratory. The project's expansion required a move to a new facility, with ninety people working on it by March 1955. Headed by Melvin Kappler and William Biel, the project used "Problem Reproducer Equipment" manufactured by Radio Corporation of America (RCA).[92]

As the project expanded from research to "production," the organization of the SRL changed as well. The core team consisted of the original participants from 1951—John Kennedy, William Biel, Robert Chapman, and Allen Newell. Kennedy, Biel, and Chapman were trained psychologists and Newell was a mathematician. Those four organized the initial efforts of the SRL and worked informally with others to organize and conduct the experiments.

From the start the members of the SRL were self-conscious of their own organization as they studied the development of organizations in their experiments. In November 1951 Chapman wrote to the group about keeping a development diary of their own organization that included not only the formal communications and minutes but also the personal journals of each member, "aimed at a frank day to day account of relations between members of the team, the extent and nature of both conflicts and organizational unities and the good and bad feelings that result from each." Although the team does not appear to have adopted it, Chapman's proposal accurately reflects the self-consciousness of the team members and the relative openness of their discussions. The openness of the group seems to have worked well for the first of couple years.[93]

By July 1953 the informal organization of this core group grew a bit strained. Kennedy proposed to give each senior staff member a group responsibility as well as an individual responsibility over the group of twenty personnel in the SRL. He noted that the original crew had some difficulty in "assimilation of new personnel" into the organization, but because of rapid expansion as the program grew in notoriety and funding they had to assimilate new people. Kennedy had some difficulty in getting his staunch individualists to work together because of the "competitive games" that they all played with each other. These included "a) Academic vs non-academic work experience, b) 'Age' vs 'youth', c) 'Working' vs 'thinking', d) Statistics vs mathematics, e) Psychology vs mathematics, f) Consulting vs research, g) Verbal facility vs strong silence, and h) Each of us against all the other members in all possible combinations over the best way to do things." Kennedy urged the core team to "keep these competitive games within bounds." Otherwise, he noted, those hurt by the competition withdrew "from the cooperative game and we lost time while the wounds are licked."[94]

Continued growth led to more formality. By September 1953 Newell began to standardize the procedures for producing the aircraft tracks used as stimuli for the experiments. In October the SRL library had to be organized in a coherent manner, and in November they standardized the forms to be used in the experiments. SRL personnel developed close relationships with Air Defense Command operating divisions and training groups to understand the details of air defense operations and to fit their training operations within ADC's operations. An ADC task group supplied information required to generate the problem scenarios, which Newell's group then converted into radar scope readouts. Other SRL/ADC groups worked out the simulation details of early warning stations, Navy picket ships, and adjacent direction centers. The increasing sophistication of the experiments and the exposure of the project to more personnel led to the creation of manuals and other documentation to pass on the core group's expertise to the growing organization. Over time the efforts became less experimental because of the increasing need to train new personnel in the SRL and elsewhere to use the new methods to train others.[95]

Through RAND's many interactions with Air Defense Command, ADC officers grew comfortable with the expertise and abilities of RAND personnel. That familiarity led ADC to introduce RAND into the SAGE program. At a conference at Western Electric on March 4, 1955, Lincoln Laboratory leaders clarified their intention to end their involvement with SAGE and its master computer program after the initial development of the system. Adaptation of the equipment and the program for the many direction centers would have to be done by some other organization, as would the tests to ensure proper installation of follow-on systems.[96]

Lincoln initially proposed that Western Electric perform the task but soon found that WE did not want it. After losing their earlier battle with Lincoln over control of SAGE, Western Electric refused responsibility for making sure the computer system and its software worked. They believed that if Lincoln wanted control of the system, they should take full responsibility for it. Reluctantly, Lincoln leaders later agreed that they must "take responsibility for getting the thing done." IBM officially stated that the task was like "bore siting" a weapon and expressed their opinion that Western Electric should do it. Privately, IBM executives did not know what they would do with more than two thousand programmers after the task was done so they turned it down. Lincoln, represented by George Valley, eventually agreed that Lincoln would have to support adaptation of the computer programs. Under the circumstances, the ADES project office, charged with SAGE installation throughout the United States and Canada, had to find other organizations to carry through these tasks. With SAGE's leaders at an impasse, the door was open for ADC to step back into the picture.[97]

Although ADC had lost its bid to develop the Michigan system, Vice Commander Frederic Smith realized that ADC still had the opportunity to influence how SAGE would operate through its programming. With Lincoln bowing out, and neither Western Electric nor IBM willing to step in, ADC could regain some influence over SAGE. ADC could tailor the programming in each direction center, relying on their connection to RAND's System Training Program.

On April 7 Smith discussed potential RAND involvement in SAGE with RAND Director Collbohm and Col. Oliver Scott, the chief of the ADES office in New York. Another meeting between RAND and ADC in Colorado Springs, on April 20–22, 1955, with IBM and Lincoln present as advisers, led to agreement that Lincoln would create the master program for the Experimental SAGE Sector and the first two operational direction centers. All further installations and modifications would be made by RAND, as would training on all SAGE direction centers for Air Force personnel. On May 14 the Air Force made its formal decision and RAND expanded its System Training Project with personnel from its numerical analysis division. The new, expanded project would be headed by Melvin Kappler and William Biel.[98]

To learn about the master program in preparation for the day when they would adapt it to different direction centers, RAND programmers began to arrive at Lincoln Laboratory in July 1955. The growth of the System Training Project led RAND to separate it into a new RAND division in September, and in December the division was renamed the System Development Division (SDD). Soon its growth began to eclipse RAND itself. Software programming, a task that others thought to be a minor and routine task, soon grew far beyond the estimates of either Lincoln or RAND personnel.[99]

Programming Crisis and Response

In March 1955 Lincoln's managers and engineers prepared to program the XD-1 prototype computer that IBM delivered to Lincoln in January 1955. Group 61 leader Robert Wieser recognized that the SAGE "master direction center computer program" required more formality than Lincoln's researchers had heretofore used:

> The work now starting on SAGE master programs will require formalities which have not been necessary in the past. We are no longer preparing programs for "our" system. It will be necessary to obtain from the Air Force formal concurrence on program specifications. Changes after concurrence will require approval, and detailed program records will have to be prepared and kept up to date.[100]

To create the master program for which his group had responsibility, Wieser redesigned his organization. The first step was to prepare operational specifications and program specifications. He assigned three sections to this task—tracking, weapons direction, and program organization—totaling twenty-seven programmers and managers. Testing would be accomplished by the twenty engineers of the test program section that had been working on the Cape Cod System. They soon would move to the Experimental SAGE System using the XD-1 prototype computer. Wieser planned that the master program would be completed in early 1956, with modifications and changes made between April and October 1956. To assist the effort, John Jacobs' systems office would dedicate some time from his six workers.[101]

Wieser and Jay Forrester realized that this staff would not be sufficient. They estimated a total of 22 engineers necessary to develop and maintain the specifications and another 40 to create the coding specifications and to perform the coding itself—a total of 62. RAND planned to build its master program group to 117 by 1957. Because programming was in its infancy, programmers could not be hired; rather, they had to be brought into the organization and trained. Jacobs agreed to look into training requirements for Lincoln, IBM, Western Electric, Bell Telephone Laboratories, and ADC, with the intent of creating a programmer training course. Lincoln management duly contracted with IBM to provide the training, eventually educating more than five hundred personnel in this new function at their XD-2 facility in Kingston, New York.[102]

Lincoln's programmers based their plans for SAGE programming primarily on their experience with the 1954 Cape Cod System. With the Whirlwind computer as the core, Lincoln engineers worked with Air Force operators and officers on tracking and interception experiments. They jointly developed better interfaces from the computer screen to the operators, and in so doing they developed a large computer program to automate many tasks.

The initial programming of the Cape Cod System was rather informal but as the program grew, the programmers eventually realized that they needed to centralize a number of functions common across many elements of code. These so-called bookkeeping programs—what we now consider part of the operating system—had been distributed across many routines. In the final Cape Cod master program, programmers led by Herb Benington placed these functions in centralized routines accessible to other routines.[103]

As the Cape Cod master program grew, programmers also realized the need for more extensive documentation. Informal communication methods did not suffice and the code itself, typically written in machine language, was virtually impenetrable to anyone except the original designer. Because the engineers who tested the program were not the same as the programmers, the programmers realized they had to document the intent of the program with operational specifications. They also needed to document the code itself with coding specifications. In the Cape Cod case, the documentation came after the development of the program, not before, but it was sufficient to aid the testers in their planning and testing.[104]

Taking a cue from Lincoln Laboratory's earlier efforts, programmers divided the program into small sections of one hundred to two thousand instructions, eventually integrating them into a final program of approximately twenty-five thousand instructions. They continued to develop utility programs to help with mundane tasks such as input and output of programs and with more difficult tasks such as debugging. The Cape Cod programmers noted that documentation of program changes was important and they created a set of standard forms for that purpose. They also developed specifications for each test, documenting them in technical information releases. Believing that SAGE would be similar to the Cape Cod System, Lincoln's managers estimated that the SAGE master program would be ready in April 1956.[105]

Through 1955 Lincoln programmers worked to recreate on the prototype computers XD-1 at Lincoln and XD-2 at IBM the capabilities that they had developed on Whirlwind. At the outset this involved creating utility programs to aid programmers. This endeavor was a somewhat controversial item because it diverted scarce programming talent and computer time to tasks not directly involved with the SAGE master program. Nonetheless, Lincoln personnel persevered because they believed that the utility routines ultimately would save a great deal of effort. The routines included the compiler, librarian, read-in, checker, and utility control routines. At the same time they developed operational and coding specifications for the master program. Those specifications required extensive coordination with the 4620th Air Wing, Bell Telephone Laboratories, Western Electric, Burroughs, and other parties external to Lincoln Laboratory.[106]

Lincoln staff also prepared for coding and testing of the master program. Engineers divided the program preparation into four areas: program organization, central bookkeeping, operational subprograms, and standby duplex operation. They planned three phases of testing—parameter testing to check out each subprogram, a test of intercommunications among subprograms, and a system test of the entire integrated master program. As Herb Benington noted, SAGE's "real-time" program probably was the first large-scale program that had to be thoroughly centralized to accomplish its functions.[107]

By September 1955 the operational specifications had been drafted and the coding group estimated the size of the master program at sixty thousand instructions using approximately one million bits of storage. The group divided that mass into thirty-five subprograms of smaller size, along with twenty data tables. Lincoln circulated the operational specifications through the Air Force, Western Electric, and ADES for concurrence, as they continued to write utility programs for the XD-1 and the memory test computer. The latter machine had been built to test core memories in the early 1950s, but by 1955 Lincoln engineers had converted it to a support computer for the SAGE effort and for other tasks at the laboratory. Although always aware of the need for simulations, with RAND now onboard Lincoln redoubled its simulation efforts for system checkout and training and created a group to work with RAND on that effort.[108]

Despite their experience with the Cape Cod System, Lincoln and RAND programmers found that the SAGE programming tasks did not move forward as quickly as expected. Approval of the operational specifications dragged on for several months until in December 1955 Lincoln froze the specifications to "permit the detailed correlation of operational specifications and programs for coding specifications and subsequent programs." In other words, Lincoln programmers had developed coding specifications and begun coding without having concurrence on the operational specifications from the Air Force. This meant that they had to cross-check between the now-frozen operational specifications and the actual code. They expected to find inconsistencies and created operational modification(s) request forms to handle them.[109]

The programming effort continued to grow in size and that expansion led to the separation of computer programming from the development of operational specifications. In early 1956 Lincoln management formed a new Group 67, the Program Production Group responsible for creating coding specifications, coding, and testing the final program. By March 1956 Group 67 had completed most of the coding specifications but clearly would not meet the April 1956 schedule for completing the coding.[110]

Finally, at the end of May 1956 Lincoln's programmers admitted to the Air Force that they had substantially underestimated the programming task and that the schedule would have to slip one year to June 1957. As Lincoln's programmers explained,

> The SAGE System introduced the complication of overlapping radars and added new functions such as automatic crosstelling, height-finding, command post, and weather and weapon status totes, totaling over 100,000 instructions. The number of personnel (Lincoln and RAND) working on the problem is nearly ten times that of any previous programming job. The unavailability of experienced programmers has meant that the whole group had to learn while doing. Most of the programs represent the first try at a program for the majority of personnel. The demand for computer time has exceeded the computer hours available.[111]

The USAF and the Culture of Innovation

The major problem was that the operational requirements that had been generated in 1955 were far too complicated. In November 1956 John Jacobs put it this way, tongue-in-cheek:

> If I'm looking for transportation, should I spend $10,000 for a Lincoln Continental with air-conditioning, or should I buy a Ford? The costs are different, but the utility is about the same. What we have been doing is going over our program design and substituting Fords for Continentals, 17-inch screens for 24-inch screens, mouton for mink, and computers for "giant brains."[112]

Only when the programmers translated the operational specifications into coding specifications did they realize what a monster they had on their hands. As Jacobs put it, "the writing of the coding specifications was a time of occupational therapy for us, and marked the end of our Lincoln Continental period. At this point, we faced up to the fact that 'there was a Ford in our future.'" To be precise, when the programmers put together the coding specifications and began actual coding, they found that the master program would require 120,000 total instructions, all of which would have to be resident in the computer at the same time. Because the computer's auxiliary drums (the equivalent of a modern hard disk drive) contained only ninety-six thousand registers, there was simply no way the program would fit. Nor would the program execute fast enough to process the radar data at its basic sweep rate. Whereas the Cape Cod program required three hundred man-months to prepare, in August 1956 the estimate for the SAGE master program was five thousand man-months.[113]

To reduce the size of the program, the programming teams had to return to the original specifications to remove anything beyond the absolute minimum required for SAGE to perform its core functions. Group 61, in charge of developing the operational specifications, "devised a way of controlling the program through the use of a hierarchy of specifications—from the Lincoln/ADC Red Book (Operational Plan) to the actual coding." Essentially they eliminated any functions that could not be derived from the operational plan.[114]

After this "code scrubbing" exercise, completed in early 1957, in which they eliminated about thirty-five instructions, programmers estimated that the master program would require eighty-five thousand instructions, and the utility programs demanded another forty thousand. Even that did not include all of the functions originally intended for SAGE. These functions would eventually be phased in with later code deliveries. On top of that, programmers coded another sixty thousand instructions for testing purposes and instrumentation. Benington noted that the supporting routines equaled or exceeded in size the actual master program. He also noted that the cost of a program of this size "can easily require $55 per instruction," which implied that the cost of the programming was comparable with

SAGE equipment in 1958 included video mappers, radar height indicators, and antenna control units.

the cost of the hardware. This result, quite unexpected at the time, highlighted the significance of programming for computer-based projects of any kind.[115]

Along with the greatly expanded estimates for the amount of code necessary, Benington pointed out the need for thorough documentation of programs of such immense size. For SAGE the documentation included "Operational specifications, Program specifications, Coding specifications, Detailed flowcharts, Coded program listings, Parameter test specifications, Assembly test specifications, System operating manuals, and Program operating manuals." Benington believed that the need for such documentation was "obvious":

> The system and its program must be learned and used by management, operational-design engineers, system-design engineers, programmers, program-test engineers, evaluation personnel, and if more than one system is maintained, on-site maintenance programmers. Each of these users has very different needs.[116]

In such a large system with its accompanying documentation, changes had the potential to wreak havoc:

> Consider the problem of revising the system once the program is operational in the field. A minor change in the operational specifications is proposed. First, the cost and effects of this change must be evaluated in terms of the program, the operators, and often, the machine. In order to make the change, several hundred revisions may be required in the specifications. If the change is approved, these documents must be changed, operating manuals revised, and the program modified and thoroughly tested. The wave of changes must be coordinated smoothly.[117]

Noting that digital computers were "often sold to management on the basis of their programmed flexibility," Benington noted that, in practice, large-scale, tightly integrated systems like SAGE were hardly flexible. Their inflexibility and the associated costs of making changes were "only a symptom of the design-coordination problem in large systems."[118]

The programming crisis led to a reorganization of the laboratory, but the problem of coordinating changes made it largely ineffectual. George Valley, who was the associate director of the laboratory, transferred Robert Wieser to Division 2 along with the Cape Cod System. Division 2, which ran the radar and communications systems, was the other important player in the SAGE system. The reorganization was intended to improve communications between Divisions 2 and 6, but it did not.

Wieser believed that the experience of the Cape Cod System was not being factored sufficiently into the SAGE design. To help transfer this experience, Division 2 chief Carl Overhage called a meeting between Divisions 2 and 6 personnel. At that meeting, Robert Everett of Division 6 stated that his group was ready to freeze the hardware specifications along with most of the software. Already facing the severe problems of software size by cutting the program and its functions substantially, they were unwilling to add anything unless Division 2 proved its necessity and impact. Since Division 2 did not have the personnel to do so, Division 6 won the turf battle but alienated Overhage, who soon would become Lincoln Laboratory's director.[119]

With the reduced scope, the increased resources, and the strict processes for documenting and modifying the program, Lincoln and RAND programmers slowly ground forward, although always at a slower pace than originally planned. By fall 1956 the master program had grown to seventy-one subprograms. While Lincoln focused on the master program, RAND geared up to hire and train four hundred programmers divided into ten teams of forty members to modify, install, and test it at the various Air Force direction centers. RAND programmers devel-

The SAGE power plant at Stewart Direction Center, New York, in the late 1950s.

oped "dynamic flow diagrams" to visualize the code and assist in training the hundreds of new hires. They soon developed diagrams for the entire master program. Another important organizational change was the realization that they would have to separate the immediate, "minimum" master program from any further changes and modifications. These later modifications would be installed in a second delivery of the master program in 1959. This allowed programmers to code to a fixed set of specifications and then make changes in a controlled fashion.[120]

By June 1957 modifications to the master program were under way at the first direction center. Lincoln had earlier established a problem-reporting procedure to help document and isolate problems, and it finally became useful as the master program came under testing. At last acknowledging the long-term significance of system coordination and control, Lincoln managers made the informal systems office into Group 68 of Division 6 in April 1957. With that, systems engineering had found a permanent place in command and control systems and in the practice of software engineering.[121]

The SAGE system itself became partially operational in June 1958 when the New York sector began functioning. In August 1958 SAGE controlled a BOMARC missile to intercept a simulated enemy bomber. Many direction centers operated eventually but a number of them, mostly in the Midwest, never were built. Untested in war, the last SAGE direction centers closed in 1983. We will never know how useful SAGE would have been in battle but it was extremely influential for peacetime research and development.[122]

Conclusion

From its inception as a general-purpose aircraft simulator to its transformation into SAGE, Project Whirlwind led the development of large-scale, real-time computing. The leader of the computer effort, Jay Forrester, stamped his painstakingly thorough and formal methods on the project to ensure high reliability and at the same time driving technology forward as rapidly as possible. The high costs of his methods several times cast serious doubts on his enterprise, but at critical moments the Cold War intervened to shower more money on the project and eventually to reward Forrester and his team with the largest, most complex computer project of the 1940s and 1950s.

When the Air Force adopted Whirlwind as its computer solution to the problem of air defense, Forrester's team had to expand its horizons well beyond the

The Whirlwind computer ended its operating lifetime on June 1, 1959.

digital circuitry of Whirlwind. Ultimately the team had to learn the technologies of radar and communications, the human factors and tactics of air defense operations, and the performance of all kinds of aircraft. Their organizational interactions grew to include other faculty at MIT and Lincoln Laboratory, IBM, Bell Telephone Laboratories, Western Electric, Burroughs, Air Defense Command, Air Materiel Command, and Air Research and Development Command. Because their geographic horizons eventually encompassed the North American continent, the erstwhile electrical engineers developed methods to program computers and coordinate across that hodge-podge of organizations and technologies.

The means to accomplish those tasks were largely organizational. Without official authority, John Jacobs developed an unobtrusive systems office to coordinate activities among Lincoln, IBM, and Air Force Cambridge Research Center. After fending off efforts by Air Research and Development Command and Western Electric to reign in their authority, Lincoln's engineers expanded their unofficial authority, although MIT set limits on Lincoln's commitment to the SAGE project. Software soon became the central integrating element of the project and grew far beyond everyone's expectations. The crises that ensued led to the recruiting of RAND to take over the long-term programming job and to the further formalization of coordination activities that eventually led to permanent establishment of the systems office.

All of this occurred in an environment in which Air Force leaders perceived that only MIT had the capability to develop the system. Air Force leaders also thought that the midgrade officer cadre lacked such expertise, and so refused to back their attempts to control Lincoln. Furthermore, senior leaders did not believe air defense was nearly as important as offensive weapons such as bombers or even ballistic missiles. They believed it simply was not worthwhile to divert too many resources or officers to that task. Left to fend for themselves, Lincoln, IBM, RAND, and Bell worked out such arrangements as they could. Over the long haul, however, this casual arrangement could not last as the Air Force slowly built up its own capabilities and the Cold War's urgency eventually slowed. When speed was no longer such an issue and cost became a larger concern the Air Force stepped in and took control of their newly built systems.

Notes

1. Memo 6M-3084-1, C. R. Wieser to All Group Leaders, Div II and VI, All Staff, Group 61, Lincoln Steering Cmte, subj: Reorganization of Group 61, Mar 25, 1955, MITRE, box 1783.

2. See the arguments in Kenneth Schaffel, *The Emerging Shield: The AF and the Evolution of Continental Air Defense, 1945–1960* (Washington, D.C.: Ofc of AF Hist, 1991).

3. Kent C. Redmond and Thomas M. Smith, *Project Whirlwind Case History,* manuscript ed (Bedford, Mass: MITRE Corporation, 1975), chapter 1, and pp 2.01–2.02; email, R. R. Everett to author, Jun 27, 1999.

4. Redmond and Smith, *Project Whirlwind,* chapter 2.

5. *Ibid.,* chapter 3.

6. Mina Rees, "The Computing Prog of the Ofc of Naval Research, 1946–1953," *Communications of the ACM* 30 (Oct 1987), p 842; Redmond and Smith, *Project Whirlwind,* pp 5.1–5.8.

7. Redmond and Smith, *Project Whirlwind,* pp 5.18–5.21.

8. Atsushi Akera, "Calculating a Natural World: Scientists, Engineers, and Computers in the US, 1937–1968," Ph.D. diss, University of Pennsylvania, 1998, pp 312–315.

9. Redmond and Smith, *Project Whirlwind,* pp 5.25–5.44; Akera, "Calculating a Natural World," p 317–318; ltr, John von Neumann to Mina Rees, Dec 10, 1947, MITRE, AC-22.

10. Redmond and Smith, *Project Whirlwind,* p 6.20; for a description of the technical problems behind the overrun, see Akera, "Calculating a Natural World," pp 320–323.

11. Redmond and Smith, *Project Whirlwind,* pp 6.21–6.33, 7.28–7.36, 8.04–8.09.

12. *Ibid.,* pp 6.02, 7.17, 9.08–9.09.

13. *Ibid.,* pp 9.10–9.17; Akera, "Calculating a Natural World," pp 349–365.

14. Redmond and Smith, *Project Whirlwind,* p 9.18.

15. Robert Buderi, *The Invention that Changed the World* (New York: Simon & Schuster, 1996), pp 182–187, 357–359; George E. Valley, Jr., "How the SAGE Development Began," *Annals of the History of Computing* 7 (1985), pp 196–198.

16. Valley, "How the SAGE Development Began," pp 198–199.

17. Valley, "How the SAGE Development Began," pp 206–207; ltr, G. E. Valley to Dr. Theodore von Kármán, SAB, HQ USAF, Nov 8, 1949, in H. W. Serig, *Project Lincoln Case History,* vol I (Bedford, Mass: AFCRC, 1952), Lincoln, U-13.050, supporting document.

18. Valley, "How the SAGE Development Began," pp 208–210.

19. Redmond and Smith, *Project Whirlwind,* pp 9.20–9.21; Valley, "How the SAGE Development Began," pp 211–212; Buderi, *Invention,* pp 371–372; progress rpt on the Air Defense Systems Engineering Cmte, SAB, to the CofS, USAF, May 1, 1950, in Serig, *Project Lincoln Case History,* vol I, p 13; ltr, Ivan A. Getting to author, Jun 18, 1999.

20. Buderi, *Invention,* pp 372–374; ltr, Gen Vandenberg to Dr. James Killian, President, MIT, Dec 15, 1950, in Serig, *Project Lincoln Case History,* vol I.

21. Buderi, *Invention,* pp 374–379; Valley, "How the SAGE Development Began," pp 213–214; ltr, Dr. James Killian to Dr. F. Wheeler Loomis, Project Charles, Feb 16, 1951, in Serig, *Project Lincoln Case History,* vol I; "Charter for the Operation of Project Lincoln," in Serig, *Project Lincoln Case History,* vol I; Kent C. Redmond and Thomas M. Smith, *From Whirlwind to MITRE: The R&D Story of the SAGE Air Defense Computer,* unpublished manuscript, MITRE, 1997, chapter 10, pp 6–7; Richard F. McMullen, *The Birth of SAGE 1951–1958,* ADC Historical Study no 33, Lincoln, accession #358587, chapter 1. The other MIT group was the so-called ZORC group, named for its main actors, Jerrold Zacharias, J. Robert Oppenheimer, I. I. Rabi, and Charles Lauritsen. It also included Lloyd Berkner. See intvw, Ivan Getting with author, Oct 30, Nov 6, 1998; see also *Semi-Annual Historical Rpt for the DCS/D, 1 Jan 1953–30 Jun 1953,* AFHRA, 140.01; Thomas P. Hughes, *Rescuing Prometheus* (New

York: Pantheon Books, 1998), pp 24–27; ltr, Ivan A. Getting to author, Jun 18, 1999. According to Getting, Zacharias insisted that Valley not run the new lab.

22. Ltr, Getting, Jun 18, 1999.

23. Memo, Lt Gen K. B. Wolfe, DCS/M, to Commanding Gen, AMC, and Commanding Gen, ARDC, subj: Emergency Mobilization of Scientists and Engineers for AF Research and Development, Feb 19, 1951, in Serig, *Project Lincoln Case History,* vol I.

24. Valley, "How the SAGE Development Began," p 214. He noted that Lincoln funding ratios approximated 10:1:1 for the AF, Army, and Navy, respectively. "Proposed General Policy for the Establishment of the Air Defense Lab (M.I.T. Contractor), ca 1951, in Serig, *Project Lincoln Case History,* vol I; see also John F. Jacobs, *The SAGE Air Defense System: A Personal History* (Bedford, Mass: MITRE Corporation, 1986), pp 87–89, for a description of the battles between Lincoln Project Ofc manager Lt Col LaMontagne and Lincoln's management. Ultimately higher-level AF officials had to intervene.

25. Ltr, Brig Gen Donald N. Yates, Dir of Research & Development, Ofc DCS/D to Dr. F. W. Loomis, Dir, Project Lincoln, Oct 4, 1951, in Serig, *Project Lincoln Case History,* vol I; ltr, N. F. Twining, Vice CofS, USAF to Lt Gen E. E. Partridge, Commanding Gen, ARDC, Dec 1, 1952, in Serig, *Project Lincoln Case History,* vol III.

26. Redmond and Smith, *Project Whirlwind,* pp 8.18–8.20.

27. Akera, "Calculating a Natural World," pp 310–311; admin memo A-42-3, Jay W. Forrester to 6345 Staff and Senior Technicians, Project Whirlwind, Servomechanisms Lab, MIT, subj: Bi-Weekly Rpts, Dec 11, 1947, rev Mar 3, 1948, MITRE, box 1782.

28. Akera, "Calculating a Natural World," pp 325–326; intvw, Robert Everett with author, Oct 1 and 13, 1998, USAF HSO.

29. Akera, "Calculating a Natural World," pp 317–318, 325.

30. Redmond and Smith, *Project Whirlwind,* pp 8.21–8.22.

31. Jay W. Forrester, "Reliability of Components," *Annals of the History of Computing* 5 (1983), pp 399–400; Henry S. Tropp, moderator, "A Perspective on SAGE: Discussion," *Annals of the History of Computing* 5 (1983), p 383.

32. See intvw, Everett, Oct 1 and 13, 1998.

33. Redmond and Smith, *Project Whirlwind,* p 5.7; Akera, "Calculating a Natural World," pp 303–309; email, Everett, Jun 27, 1999. Everett stated that the word length was always sixteen bits. Akera gave a good description of the early problems with the design, but in my opinion gave a bit too negative a twist to the tale. He noted that adoption of the von Neumann IAS-style architecture entailed a series of "commitments" even when design knowledge was uncertain, and that those led to substantial overruns. Akera noted later that there might have been no better alternatives. I think this process is normal because as a designer one must start with something, rather than nothing, to work out the implications of the design. The implications, as Akera rightly noted, are uncertain at the time a top-level design is selected. The only alternative is to develop more than one design at the same time and work out *their* implications—an even more expensive proposition.

34. Akera, "Calculating a Natural World," pp 360–361.

35. *Ibid.*, pp 315–316, 322–323; email, Everett, Jun 27, 1999. Everett stated that Forrester's insistence on easy maintenance was critical, leading to the "boxy" mechanical design. That contradicts Akera's statement that Sylvania used its standard packaging and that MIT's engineers did not pay much attention to structural design.

36. Redmond and Smith, *Project Whirlwind,* pp 5.13–5.14, 8.17.

37. Air Defense Systems Engineering Cmte, "The Air Defense System," Oct 24, 1951, in Serig, *Project Lincoln Case History*, vol I.

38. *Ibid.*

39. Brief Summ of Activity Planned by Project Lincoln During F/53, Jan 9, 1952, in Serig, *Project Lincoln Case History*, vol II; organization chart, Lincoln Lab, MIT, Aug 1, 1952, MITRE.

40. See Emerson W. Pugh, *Memories That Shaped an Industry* (Cambridge: MIT Press, 1984); Valley, "How the SAGE Development Began," pp 216–217.

41. Morton M. Astrahan and John F. Jacobs, "History of the Design of the SAGE Computer—The AN/FSQ-7," *Annals of the History of Computing* 5 (Oct 1983), pp 343–344; Emerson W. Pugh, *Building IBM: Shaping an Industry and Its Technology* (Cambridge: MIT Press, 1995), pp 207–210.

42. Extract from Lincoln Lab technical memo no 20, 2d draft, Jan 2, 1953, in Serig, *Project Lincoln Case History*, vol III.

43. Pugh, *Building IBM*, pp 210–213.

44. Memo M-1891, A. P. Kromer to J. W. Forrester et al., subj: Summ of IBM–MIT Collaboration, Feb 1, 1953 to Feb 23, 1953 inclusive, Mar 9, 1953, p 2, quoted in Redmond and Smith, *From Whirlwind to MITRE*, chapter 17, p 4; memo M-1696, Harris Fahnestock to All Staff Members, Digital Computer Lab, subj: Liaison with IBM Corporation, Oct 28, 1952, MITRE, Redmond and Smith Papers, box 15, ASC-1783; Astrahan and Jacobs, "History of the Design," pp 344–345; Redmond and Smith, *From Whirlwind to MITRE*, chapter 19, pp 3–5.

45. Astrahan and Jacobs, "History of the Design" pp 344–346; Jacobs, *SAGE Air Defense System*, pp 55–62.

46. Jacobs, *SAGE Air Defense System*, pp 62–65; Astrahan and Jacobs, "History of the Design," pp 346–347.

47. Jacobs, *SAGE Air Defense System*, p 67.

48. *Ibid.*, p 65; Astrahan and Jacobs, "History of the Design," pp 346–347.

49. Jacobs, *SAGE Air Defense System*, p 67; Redmond and Smith, *From Whirlwind to MITRE*, chapter 19, p 10; file memo, C. F. Lynch, R. P. Crago, and J. W. Forrester, subj: Joint IBM–MIT Approvals, Jan 26, 1954, MITRE, box 1785.

50. Jacobs, *SAGE Air Defense System*, p 68.

51. Admin memo A-111, Jay W. Forrester to All Personnel, Electronic Computer Div, Servomechanisms Lab, MIT, subj: Staff Organization, Dec 11, 1950, MITRE, Redmond and Smith Papers, box 15, ASC-1783.

52. Redmond and Smith, *Project Whirlwind*, pp 11–26.

53. Memo M-2086, C. R. Wieser and W. G. Welchman to 6673 Group, subj: Temporary Allocation of Responsibilities in Project 6673, Feb 15, 1951, quoted in Redmond and Smith, *From Whirlwind to MITRE*.

54. Memo M-1284, John Carr and John T. Gilmore, Jr., to the Applications Group, Digital Computer Lab, subj: Method of Preparing Subroutines for the Subroutine Library, Sep 24, 1951, MITRE, box 1776; memo M-1350, J. T. Gilmore to the Mathematics Group, Digital Computer Lab, subj: Operational Procedures on the Whirlwind Computer, Dec 10, 1951, MITRE, box 1776.

55. Quarterly Progress Rpt, Div 6-Digital Computer, Lincoln Lab, MIT, Sep 15, 1953, MITRE, pp 43, 47.

56. *Ibid.*, pp 35, 47.

57. Memo, Col R. M. Osgood, Dir of Electronics Ofc, Dep for Development, HQ ARDC, to Commanding Gen, AFCRC, subj: Command Policy as Regards Lincoln Transition System, Feb 20, 1953, in Serig, *Project Lincoln Case History*, vol III; memo, Milton A. Chaffee, Dep Dir for Systems, to CRR, Dr. E. G. Schneider, subj: Requirements for Setting up Air Defense System Project, Mar 23, 1953, in Serig, *Project Lincoln Case History*, vol III. Blasingame would soon

transfer to Schriever's group in El Segundo to work on ICBMs.

58. Memo, Col R. M. Osgood, Dir of Electronics Ofc, HQ ARDC, to Commanding Gen, AMC, subj: Phasing Lincoln Lab Developments into Production, Mar 30, 1953, in Serig, *Project Lincoln Case History,* vol III.

59. McMullen, *Birth of SAGE,* pp 10–12.

60. *Ibid.,* p 11; Paul N. Edwards, *The Closed World: Computers and the Politics of Discourse in Cold War America* (Cambridge: MIT Press, 1996), pp 96–97.

61. Valley, "How the SAGE Development Began," pp 221–224.

62. *Ibid.,* pp 221–224; Edwards, *Closed World,* pp 96–97; see Getting, in Jacob Neufeld, ed, *Reflections on R&D in the USAF: An Interview with Gen Bernard A. Schriever and Gens Samuel C. Phillips, Robert T. Marsh, and James H. Doolittle, and Dr. Ivan A. Getting,* conducted by Dr. Richard Kohn (Washington, D.C.: Ctr for AF Hist, 1993), p 51.

63. Ltr, James R. Killian, Jr., to Secretary of the AF Thomas J. Finletter, Jan 9, 1953, in Serig, *Project Lincoln Case History,* vol III.

64. Redmond and Smith, *From Whirlwind to MITRE,* chapter 11, pp 12–13.

65. Ltr, Thomas J. Finletter to President Killian, Jan 15, 1953, in Serig, *Project Lincoln Case History,* vol III.

66. Ltr, Lt Gen E. E. Partridge, Commander ARDC, to Dr. James R. Killian, Jan 28, 1953, in Serig, *Project Lincoln Case History,* vol III.

67. Ltr, Maj Gen D. L. Putt, Vice Commander, ARDC, to Maj Gen Raymond C. Maude, AFCRC, Mar 16, 1953, in Serig, *Project Lincoln Case History,* vol III; ltr, Horace S. Ford, Dir, Div of Defense Labs, MIT, to Donald L. Putt, HQ ARDC, Mar 4, 1953, in Serig, *Project Lincoln Case History,* vol III; ltr, Maj Gen R. C. Maude to Maj Gen D. L. Putt, Feb 26, 1953, in Serig, *Project Lincoln Case History,* vol III.

68. Memo, Maj Gen D. L. Putt, Vice Commander, ARDC, to Commanding Gen, AFCRC, subj: Revision of Command Policy Pertaining to Lincoln Transition System, May 6, 1953, in Serig, *Project Lincoln Case History,* vol III; see also ltr, Lt Gen E. E. Partridge, Commander ARDC, to Dr. James R. Killian, May 6, 1953, in Serig, *Project Lincoln Case History,* vol III; and ltr, Lt Gen E. E. Partridge, Commander ARDC, to Dr. Harlan Hatcher, President, University of Michigan, May 6, 1953, in Serig, *Project Lincoln Case History,* vol III; McMullen, *Birth of SAGE,* pp 12–13; Forrester, in Tropp, "Perspective on SAGE," p 381.

69. Memo, Col R. M. Osgood, Dir of Electronics Ofc, Dep for Development, ARDC, to Commanding Gen, AFCRC, subj: Air Defense Coordination with the Dept of the Army, May 15, 1953, in Serig, *Project Lincoln Case History,* vol III; ltr, G. E. Valley, Asst Dir, MIT Lincoln Lab, to Maj Gen Raymond C. Maude, AFCRC, Jun 12, 1953, in Serig, *Project Lincoln Case History,* vol III.

70. Memo, Milton A. Chaffee, Air Defense Systems Project, Systems Project Ofc, AFCRC, to distribution list, subj: Mtg on F–102 Weapon System 28 and 29 Jul 1953, Jul 23. 1953, in Serig, *Project Lincoln Case History,* vol III. In late Jul, AFCRC and Lincoln personnel met with F–102 representatives, including Dr. Rube Mettler and John Rubel from Hughes Aircraft. Mettler soon would become a leader in the AF's ICBM progs, and Rubel later implemented important organizational reforms as Robert McNamara's right-hand man in the OSD.

71. Memo M-2315, R. A. Nelson, Memo M-2315, MIT Digital Computer Lab, to J. W. Forrester, subj: Mtg on Preparation of Material for Colorado Springs 17 Aug Mtg, Jul 22, 1953, in Serig, *Project Lincoln Case History,* vol III.

72. Jacobs, *SAGE Air Defense System,* pp 82–83; Valley, "How the SAGE Development Began," pp 224–225.

73, Jacobs, *SAGE Air Defense System,* pp 83–84.

74. Memo, Col Richard H. Curtis, DCS/M, AFCRC, to Commander, Rome AF Depot, Griffiss AFB, subj: Recommended Manufacturer for AN/FSQ-7, Aug 12, 1953, in Serig, *Project Lincoln Case History,* vol III.

75. Memo, Col Richard H. Curtis, DCS/M AFCRC, to Commander, AMC, attn: Industrial Facilities Ofc, subj: Facilities Requirements—IBM Corporation, Aug 7, 1953, in Serig, *Project Lincoln Case History,* vol III.

76. McMullen, *Birth of SAGE*, pp 21–30; Jacobs, *SAGE Air Defense System,* p 85.

77. Redmond and Smith, *From Whirlwind to SAGE,* chapter 24.

78. Minutes of Conference Regarding Responsibilities for Transition System, Jul 2, 1954, at MIT Lincoln Lab, "Div III (Valley, G. E., Hist of SAGE)" file.

79. C. Robert Wieser, "The Cape Cod System," *Annals of the History of Computing* 5 (Oct 1983), pp 362–369; minutes, Lincoln Steering Cmte Mtg, Jul 6, 1954, MITRE, AC-22; memo 6M-3084, Jay. W. Forrester, subj: Group 61 Reorganization, Oct 13, 1954, MITRE, box 1783; memo 6L-173, Jay W. Forrester to Lincoln Steering Cmte, subj: Organization and Tasks of Div 6, 16 Nov 1954, MITRE, box 1782; memo, Dir's Ofc, Lincoln Lab, to All Staff Members, subj: Lincoln Production Coordination Ofc—Admin Bull No [unreadable], Jun 24, 1954, Lincoln, "SAGE Production Coordination Ofc (Dr. Valley)" file.

80. Memo L-154, A. P. Kromer to R. R. Everett and Jay W. Forrester, Jun 11, 1954, Lincoln, "Div. III (Valley, G. E. Hist of SAGE)" file; memo, Frank Heart to C. R. Wieser, subj: Visit to Lab by RAND Corporation Personnel, Dec 29, 1954, MITRE, box 1782.

81. J. L. Kennedy, "RAND Extension of Bavelas Project," in M. M. Flood, ed, "Rpt of a Seminar on Organizational Science," RAND Research Memo RM-709, Oct 29, 1951.

82. Steve J. Heims, *The Cybernetics Group* (Cambridge: MIT Press, 1991), pp 220–221; see intvw, Herbert Simon with author, Jul 1 and 2, 1998.

83. Kennedy, "RAND Extension"; ltr L-12381, Allen Newell to Dr. Freed Bales, Oct 16, 1951, CMU, Allen Newell Papers, "Newell, Allen—RAND—511—Notes—1952," box 70; memo M-4206, J. L. Kennedy to H. Speier, subj: Progress Rpt No 1 on the SRL, Oct 26, 1951, "1951–1953 SRL (Rand)" folder 23, box 21, SDCP, CBI 90; F. N. Marzocco, "The Story of SDD," SD-1094, Oct 1, 1956, p 1, "RAND Corp. SDD Hist 1956–7" folder, box 1, SDCP, Burroughs Collection, CBI 90.

84. R. L. Chapman, with assistance from W. C. Biel, J. L. Kennedy, and A. Newell, "The SRL and Its Program," Jan 7, 1952, RAND D-1166.

85. *Ibid.*, pp 17–21.

86. Chapman, "SRL," pp 13–17; Allen Newell, "Description of Experiments, II. The Task Environment," Jul 29, 1955, RAND P-659; J. L. Kennedy (likely author), "The Research Pay Offs," ca Oct 1951, "1951–1953 SRL (Rand)" folder 23, box 21, SDCP, CBI 90; memo M-64, John L. Kennedy to Hans Speier, subj: Progress Rpt No 3 on the SRL, Jan 7, 1952, "1951–1953 SRL (Rand)" folder 23, box 21, SDCP, CBI 90.

87. W. C. Biel, R. L. Chapman, J. L. Kennedy, A. Newell, "The SRL's Air Defense Experiments," Oct 23, 1957, RAND P-1202, pp 20–22.

88. Marzocco, "Story of SDD," p 1; ltr, F. R. Collbohm to Maj Gen Frederic H. Smith, Jr., Vice Commander, ADC, May 27, 1952, "SRL 1952–1957" folder 24, box 21, SDCP, CBI 90.

89. *Ibid.*

90. Marzocco, "Story of SDD," pp 1–2; W. C. Biel et al, "SRL's Air Defense Experiments," pp 8–10.

91. Marzocco, "Story of SDD," pp 2–4.

92. *Ibid.*, pp 4–5.

93. Memo, Robert L. Chapman to John L. Kennedy, "Dear Diary," Nov 12,

1951, "1951–1953 SRL (RAND) folder 23, box 21, SDCP, CBI 90.

94. Draft memo, John L. Kennedy to SRL Senior Staff, subj: Organization of Psychology Research Labs, Jul 16, 1953, "1951–1953 SRL (RAND)" folder 23, box 21, SDCP, CBI 90.

95. General memo, A. Newell, subj: General Admin Procedures, 9/29/53, "COBRA 1953–1954" folder 14, box 16, SDCP, CBI 90; memo M4358, SRL Staff to J. L. Kennedy, subj: Activities of the SRL: Sep 1953, Oct 2, 1953, "SRL Rpts" folder 22, box 21, SDCP, CBI 90; see also activity rpts with same title for Oct and Nov; System Training Prog Staff, The System Training Prog for the ADC (U), RAND RM-1157-AD, Nov 3, 1953, pp 16–18; Cogwheel Manual, "Cogwheel Manual 1954" folder 16, box 16, SDCP, CBI 90.

96. Ltr, Jay W. Forrester to Col O. M. Scott, ADES Project Ofc, subj: Preparation Computer Progs SAGE System, Mar 25, 1955, "SAGE System Training Prog (2 of 2) 1955–1958" folder 22, box 19, SDCP, CBI 90; memo 6L-189, Jay W. Forrester, Robert Everett, C. Robert Wieser to Lincoln Steering Cmte, subj: A Proposal for Accomplishing the SAGE System Computer Programming Tasks Outlined in Memo 6M-3416, Mar 11, 1955, MITRE, box 1778; draft memo 6L-190, Jay W. Forrester, Robert R. Everett, C. Robert Wieser, to Div 6, Lincoln Lab, subj: Proposed Responsibilities for Tasks in Memo 6M-3416, prepared for conference at WE, Mar 17, 1955, Mar 16, 1955, MITRE, box AC-1778.

97. Marzocco, "Story of SDD," p 7; ltr, S. P. Schwartz, Asst Project Manager, Logistics, WE, to Col O. M. Scott, Dep Chief, ADES Project Ofc, subj: Computer Programming, Apr 1, 1955, "SAGE System Training Prog (2 of 2) 1955–1958" folder 22, box 19, SDCP, CBI 90; ltr, J. E. Zollinger, IBM, to Col O. M. Scott, ADES Joint Project Ofc, subj: Proposal for IBM Responsibilities and Manpower Requirements for SAGE System Programming Tasks, Apr 5, 1955,

"SAGE System Training Prog (2 of 2) 1955–1958" folder 22, box 19, SDCP, CBI 90; ltr, Col O. M. Scott to Commander, ADC, attn: Col O. T. Halley, Jr., subj: Computer Programming, Apr 7, 1955, "SAGE System Training Prog (2 of 2) 1955–1958" folder 22, box 19, SDCP, CBI 90; Tropp, "Perspective on SAGE," p 386; minutes of computer programming mtg held on Mar 17, 1955, at ADES Project Ofc, MITRE, box 1782; memo, G. E. Valley to J. W. Forrester, subj: Notes on Final Portion of Mtg at 220 Church Street, New York, of 17 Mar, Mar 18, 1955, MITRE, box 1782.

98. Ltr, Col Oliver Scott, ADES Acting Chief, to Commander, ADC, subj: Computer Programming, Apr 7, 1955, "SAGE System Training Prog (2 of 2) 1955–1958" folder 22, box 19, SDCP, CBI 90; Marzocco, "Story of SDD," p 7; memo, Paul Armer to Numerical Analysis Div, subj: Admin Changes, May 6, 1955, "SAGE 1957–1965" folder 19, box 19, SDCP, CBI 90.

99. Author unknown, "Outline of SDC History (I–IV)," Jun 15, 1964, "SDC Hist (Including STP, PACOM, Command Research) 1960–1964" folder, SDC series, box 1, Burroughs Collection, CBI 90; Marzocco, "Story of SDD," pp 7, 8; Claude Baum, *The System Builders: The Story of SDC* (Santa Monica: SDC, 1981).

100. Memo 6M-3084-1, C. R. Wieser to All Group Leaders, Div II and VI, All Staff, Group 61, Lincoln Steering Cmte, H. Sherman, subj: Reorganization of Group 61, Mar 25, 1955, MITRE, box 1783.

101. Memo, H. W. Fitzpatrick to N. Mcl., SAGE, subj: IBM, WE and AF Personnel—Temporary Assignment to Project Lincoln, Mar 3, 1955, "SAGE–IBM" file, Lincoln.

102. Memo 6L-189, Forrester, Everett, and Wieser, Mar 11, 1955; ltr, Jay W. Forrester to J. E. Zollinger, IBM Corporation, subj: Implementation of Programmer Training for SAGE, Apr 28, 1955; memo 6M-3713, J. F. Jacobs to

Jay W. Forrester and distribution list, subj: RAND–Lincoln Relationship, Jul 7, 1955, MITRE, box 1782; Jacobs, *SAGE Air Defense System,* pp 105–107; memo, H. D. Benington to P. R. Bagley, subj: IBM Programmer Training, Oct 27, 1955, MITRE, box 1782.

103. Quarterly Progress Rpt, Div 6-Digital Computer, Lincoln Lab, MIT, Mar 15, 1955, MITRE, p 5.

104. *Ibid.,* p 6.

105. *Ibid.,* pp 7–8; Quarterly Progress Rpt, Div 6-Digital Computer, Lincoln Lab, Jun 15, 1955, MITRE, pp 4, 17.

106. Quarterly Progress Rpt, Jun 15, 1955, pp 4, 17; Quarterly Progress Rpt, Sep 15, 1955, p 12.

107. Quarterly Progress Rpt, Jun 15, 1955, pp 10–13; Herbert D. Benington, "Production of Large Computer Programs," *Annals of the History of Computing* 5 (1983), p 356.

108. Quarterly Progress Rpt, Jun 15, 1955, pp 35–36; Quarterly Progress Rpt, Sep 15, 1955, p 3.

109. Quarterly Progress Rpt, Mar 15, 1956, p 73.

110. *Ibid.*

111. Quarterly Progress Rpt, Jun 15, 1956, p 69.

112. Jacobs, *SAGE Air Defense System,* pp 180–181.

113. *Ibid.,* pp 111, 197, 259; ADES Project Ofc, Review and Evaluation of SAGE Project, Aug 6, 1956, MITRE, box 1782, pp 20–28.

114. Jacobs, *SAGE Air Defense System,* pp 109–110.

115. Quarterly Progress Rpt, Jun 15, 1956, p 70; Benington, "Production of Large Computer Programs," pp 357–359; Quarterly Progress Rpt, Mar 15, 1957, p 69.

116. Benington, "Production of Large Computer Programs," pp 360–361.

117. *Ibid.*

118. *Ibid.*

119. Jacobs, *SAGE Air Defense System,* pp 111–113.

120. Quarterly Progress Rpt, Sep 15, 1956, pp 81–82; Quarterly Progress Rpt, Div 6-Digital Computer, Lincoln Lab, MIT, Dec 15, 1956, pp 66–67; memo 6M-4600, R. P. Mayer and O. E. Ellingson to Group 66 Staff (Lincoln and RAND), subj: Group 66 Training Document: Dynamic Flow Diagrams for SAGE, Nov 15, 1956, MITRE, box 1782; intvw, Robert Everett with author, Oct 13, 1998, USAF HSO.

121. Quarterly Progress Rpt, Div 6-Digital Computer, Lincoln Lab, MIT, Jun 15, 1957, MITRE, pp 71, 87; memo 6M-3883, E. F. Ennis, BTL, W. J. Mitchell, IBM, C. W. Uskavitch, Group 22, C. W. Watt, Group 64, to E. S. Rich, subj: Rpt of Cmte Studying Failure Reporting in the SAGE and Experimental Subsectors, Sep 9, 1955, MITRE, box 1773; memo 6M-3904, from C. W. Watt to E. S. Rich, subj: Minutes of Mtg to Discuss the Recommendations of the Cmte on Failure Reporting, Sep 26, 1955, MITRE, box 1773; memo 6A-207, R. R. Everett to distribution list, subj: Group 68, Apr 11, 1957, MITRE, AC-8.

122. Hughes, *Rescuing Prometheus,* p 65; Schaffel, *Emerging Shield,* p 207; Edwards, *Closed World,* pp 108–109.

Chapter 5

Standardizing the Systems Approach

"Is it true that a knight of yours killed a dragon with his bare hands?" the prince asked.

"Quite true," replied Arthur.

"Surely that knight must rank first among all the knights of the Round Table," the prince opined.

"Not so," said the King. "In fact, that knight ranks 173rd. The knights who rank foremost among the knights of the Round Table are those who write the guide books for dragon slaying. As a matter of fact, the guide books say that you are supposed to slay a dragon with a sword, so killing one with bare hands shows an improper understanding of the correct procedure. . . ."

Ed Bensley, engineer, Lincoln Division 6, late 1950s.[1]

The scientists, engineers, and Air Force officers involved in ballistic missile projects or in command and control system projects developed new processes to coordinate and manage their respective complex technologies. Personnel developing each defense approach knew of the existence of the other and a few individuals had worked in both venues, but for the most part they developed independently. Remarkably, the methods created for ballistic missiles and C^2 systems resembled each other in a number of ways; the technical considerations underlying both approaches drove the two toward similar methods.

Both ballistic missiles and semi-automatic air defenses like SAGE were extraordinarily complex state-of-the-art technologies in the 1940s and 1950s. The complexity was wide, requiring knowledge of many disciplines, and deep, typically demanding several years of technical education in each of the disciplines. In addition to pushing technologies to their limits, these systems required the utmost reliability. In the case of ballistic missiles, this was necessary simply to get them to fly; in the case of SAGE, the computing system had to operate continuously for years when other computer systems of that time failed every few hours.

Because the Air Force lacked its own internal R&D capabilities and had few technically trained officers, it relied on external organizations to provide the necessary competence. For ballistic missiles, Ramo-Wooldridge supplied the necessary integrative functions, and for SAGE, MIT's Lincoln Laboratory played the same role. As the numbers of technically trained officers increased through training programs and recruiting, however, more officers criticized the inordinate power of R-W and Lincoln because the officers believed the Air Force could do as good a job or better.

The roles of R-W and Lincoln diminished when ballistic missiles and SAGE became operational, but new developments in space systems counteracted that trend. Both civilian organizations diversified because communications and monitoring systems were needed to communicate with and locate spacecraft. The Sputnik crisis created new urgent demands in the late 1950s, but the growing trend in the 1960s was to emphasize cost control at the expense of schedules.

The result of these technical and organizational trends in the Air Force was the development of new standards for R&D, applicable across the service and its contractors. Promoted also by Robert McNamara, John F. Kennedy's secretary of defense after 1961, these new standards would slow the pace of development and would replace ad hoc relationships between the military and the technical experts with formal relationships more typical of government–industry relations. Although the government pressed for most of these developments, scientists advocated some of the changes as they pressed to separate the increasing formality of technology development from their cherished freedom to perform research.

The Researchers' Refusal: The Formation of SDC and MITRE

By 1956 the growth of RAND's System Development Division, dedicated to SAGE programming and officer training, began to eclipse RAND itself. The SDD's work became increasingly routine as its personnel developed standardized training courses for hundreds of Air Force officers and trained hundreds of programmers to modify the SAGE master program for each air defense sector. When these tasks were completed SDD would maintain the programs and continue to train operators.

The work involved did not fit RAND's charter of basic and applied research. As the System Training Program and SAGE programming grew, RAND's management became increasingly uncomfortable with having those large development tasks housed within their research organization. They had taken on the SAGE programming task as a favor to the Air Force because no commercial organization wanted to do it. The continued growth of SDD required separate facilities and its huge size demanded that RAND's management give it increasing attention. By 1956, with the concurrence of the Air Force, RAND's board of trustees decided to separate SDD from RAND. The articles of incorporation for the new nonprofit System Development Corporation (SDC) were filed in November 1956, and in December 1957 SDC began formal operations. SDC was similar to RAND in a number of ways, particularly in its close ties to the Air Force, and like RAND, SDC's charter did not explicitly require it to contract exclusively with the Air Force.[2]

Across the country, MIT managers struggled with the same issue. As Lincoln Laboratory grew in size and its work on SAGE moved from research to development and operations, MIT and Lincoln executive management worried that SAGE would distort MIT's academic orientation away from research. Furthermore, if MIT took on development tasks, industry might view MIT as a competitor and that would alienate them and reduce potential grants and corporate philanthropy.[3] On the one hand, MIT had built the laboratory to aid with the national defense emergency and it wanted to ensure SAGE's success. On the other hand, SAGE research essentially was complete. Lincoln's management agreed that they ought to withdraw from SAGE as quickly as possible so the question became what to do next.[4] There were new research opportunities, particularly in space projects, computing, and communications. Computer programming was the first area in which MIT management had limited Lincoln's "development" ambitions, thus leading to RAND's selection as the programming contractor for SAGE master program modification and installation. The second limiting factor was the problem of "integration."

At the start of the SAGE project, the AN/FSQ-7 computer was to take signals from land-based radar sites, process the data, and direct a specific class of interceptors to the target location. Between 1952 and 1957, however, several new radar systems and defensive weapons were developed and Lincoln had to integrate them into the air defenses. These included Texas Towers (offshore radar systems placed on modified oil drilling rigs off the New England coast), Navy picket ships, airborne early warning aircraft, new interceptors such as the F–106, the Air Force's BOMARC ground-to-air missiles, and the Army's Nike ground-to-air missile. This involved changes to SAGE software, and hardware changes to convert the data into different formats. The F–106 and the Army's Nike each had its own computer system and SAGE would have to talk with those computers to ensure success.[5]

To complicate matters, integration also involved the Army, Navy, and their electronic and weapon systems contractors. Neither service was particularly eager to have its systems controlled by the Air Force's new computer. The Army, in particular, was firmly set against Air Force control of Nike and argued for local control in case of failures in the SAGE system. Local control of Nike conveniently would guarantee Army control of their weapon as well.[6]

Lincoln was at the center of these disputes because it had to ensure the master computer program could work with the other systems. In January 1956 Jay

Forrester described Division 6 procedures for integrating new weapons. Stating that Lincoln's personnel were already "fully committed" to SAGE, Forrester planned to have each "weapons system contractor" carry out the integration with Lincoln's assistance. Lincoln would "make a small group available for broad coordinating" of these weapons, and would, from time to time, "operate familiarization courses" on SAGE for contractor personnel.[7]

Lincoln's commitment to withdrawal from SAGE concerned senior Air Force leaders. Maj. Gen. Frederic Smith of Air Defense Command became alarmed that Air Research and Development Command might select RAND's System Development Division to take on the master computer programming. He feared that "this would overburden RAND and measurably detract from the output ADC expects from then [*sic*] in making up detailed specific area programs." Smith believed that if Lincoln did not want to take on this job, ARDC should force them to by threatening to "take away the XD-1 computer and give the program to someone willing to do a complete R&D job and not selected parts."[8]

The Air Force BOMARC ground-to-air missile during launch. Eventually it was controlled by SAGE.

The F–106 all-weather manned interceptor was controlled by the SAGE air defense system.

Others close to the program agreed. In December 1955 Col. Albert Shiely, the head of the Air Defense Systems Operating Division, a division of the Air Defense Engineering Services office, described the situation in the following terms:

> It is our feeling that the Lincoln Laboratory should be prevented from undertaking any new programs until it has discharged fully its responsibility in the SAGE Program. Since it is not possible to determine what Lincoln's responsibility really is, we have listed below some unfinished business that Lincoln must accomplish or agree that the Air Force (AFCRC) can accomplish without Lincoln's obstruction.[9]

Shiely then listed a number of issues, the first of which was the integration of BOMARC, the F–102A, and other weapons into SAGE. Lincoln associate director George Valley had objected to the Air Force Cambridge Research Center Lincoln project office having "System Engineering responsibility," stating that it belonged to Lincoln. As Shiely astutely noted, "Lincoln, on one hand, states that its objective is to get out of the SAGE program, and, on the other hand, actively obstructs any effort on the part of AFCRC to participate or assume any responsibility. This situation is untenable."[10]

The problems of coordinating among the Army, Navy, Air Force, and contractors without any group having authority also concerned officers at the highest levels. In March 1956 ARDC established an Electronic Supporting Systems Division within the Directorate of Systems Management to help coordinate efforts and to seek potential contractors to assist Lincoln. That same month, ARDC recognized Lincoln's determination to withdraw from SAGE and stated that some of Lincoln's responsibilities would have to be shouldered by AFCRC, Rome Air Development Center, Wright Air Development Center, or RAND.[11]

The problem was not simply a question of integrating some technologies into SAGE but rather a larger question of "Air Defense Systems Engineering," according to Colonel Shiely in the ADES office. Shiely wrote to George Valley that the Air Force believed strongly that Lincoln needed to assume this effort. Concurrently, MIT management persuaded Valley that Lincoln should not undertake further SAGE tasks. In the resulting impasse, ARDC requested that Lincoln create a plan to recommend some organization to perform the overall systems engineering, while it handled integration issues on a case-by-case basis.[12]

Concerned that without Lincoln's official support no plan would work, the Air Force felt it needed officially to offer Lincoln the systems engineering job. It did so in May 1956 when the Electronic Supporting Systems Division officially tendered the offer. After some comments and discussion that summer, Lincoln officially declined the job on September 24, 1956. Lincoln managers favored Lincoln's traditional independence in research instead of the increased Air Force control that would come with increased size and a turn toward development. They also were concerned with the Air Force's management. With numerous weapons, contractors, and government organizations involved, the Air Force had not developed "any semblance of overall coordination except at the very top."[13]

With Lincoln officially out of the running, ARDC turned next to industry to supply its system engineering needs. ARDC asked thirteen contractors to submit proposals at a meeting at ARDC Headquarters on October 23. Only RCA did so and ARDC selected the firm to perform the task. ARDC submitted its selection to Air Force Headquarters for approval, but Headquarters promptly quashed the choice. Although the reasons for that are unclear the problems with a profit-making corporation performing systems engineering were becoming obvious in Schriever's use of Ramo-Wooldridge on the West Coast and it is likely that Air Force leaders were not about to create a similar situation with SAGE. Furthermore, the hardware ban that went along with the systems engineering job discouraged others from bidding: the systems engineering contractor would not be allowed to bid for production contracts because of its "insider" knowledge gained through working with the Air Force.[14]

This left only the SDC, which wanted the job but really was not qualified. SDC president Melvin Kappler argued that his organization should become the systems engineering contractor because his personnel had substantial experience with air defense simulation and programming. He realized that SDC needed to find work for its hundreds of programmers after SAGE was complete and he believed that SAGE systems engineering would serve that purpose. Most of the Air Force's leaders favored working with Lincoln and MIT, however, because they essentially had been doing this task all along. After an inconclusive meeting among SDC, Lincoln, and Air Force representatives in July 1957 during which SDC and Lincoln presented their ideas, Lincoln and Air Force experts came away unimpressed with SDC ideas. It was apparent that SDC considered systems engineering to be little more than operations analysis and computer programming.[15]

Not everyone at Lincoln was against Lincoln's involvement with systems engineering. Robert Everett had become chief of Division 6 following Jay Forrester's departure in June 1956. Just prior to the SDC and Lincoln presentations, spurred by an Air Force proposal to coordinate its efforts more successfully, he proposed that Lincoln assume systems engineering responsibility for SAGE and future systems.[16] He reasoned that Division 6 had been doing much of the weapon systems integration anyway, and already had committed to revising the basic SAGE program in 1959. More important, his division had deeply influenced air defense work since 1950 and he wanted to maintain that level of influence. If Lincoln concentrated only on research, it would lose its substantial power over events. According to Everett, "there is a strong feeling that the research and development effort cannot vigorously influence systems decisions if separated from the systems work. Equally, systems work becomes stagnant without close support of a vigorous research and development effort." To assume the systems engineering role, Lincoln would have to take on only one additional task: the maintenance of a single master specification for the entire system.[17]

In August 1957 Lincoln director Carl Overhage proposed that the laboratory do the job for the Air Force as described in Everett's proposal, with the proviso

James McCormack, Jr., retired from the Air Force with the rank of general and became MIT's vice president for industrial relations. He originally was slated to head the Air Force's ICBM programs but health problems intervened. An early analyst of the Air Force's R&D system, McCormack saw Bernard Schriever take on the radical new missile programs.

Lincoln Laboratory in 1958.

MITRE's first offices were in the ADSID buildings at Lincoln Laboratory.

that Lincoln maintain its customary independence. By this time, however, ARDC was searching for an industrial contractor that would be easier to control than were Lincoln and MIT. When that search effort failed, ARDC was left where it had started, bargaining with MIT to figure out how systems engineering would be performed in the future. The Air Force was in a bind: Air Force leaders had disallowed those few profit-making companies that were interested in the job, such as RCA. Bell Laboratories and Western Electric did not want it. That left MIT, whose administrators wanted out even if some at Lincoln did not, and SDC, which wanted the job but was not qualified.[18]

Despite Everett's enthusiasm, MIT management wanted to distance itself from large-scale development. Nonetheless, they did not want to see the SAGE system fail or their reputations ruined by abandoning their duty to see SAGE through to completion. MIT acting president Julius A. Stratton and Lincoln director Overhage agreed that Lincoln would work with the Air Force to find an acceptable solution. That task fell to recently retired Air Force Gen. James McCormack, Jr., who had just become MIT's vice president for industrial relations. McCormack was well suited to the task for his experience included formation of the nonprofit Institute for Defense Analyses and the nonprofit arm of Western Electric, the Sandia Corporation.

With MIT's original contract for Lincoln's services ending on January 1, 1959, MIT's administrators had good reason to negotiate a favorable new agreement with the Air Force. In view of the eagerness of Division 6 to take on SAGE integration and systems engineering, and the reluctance of the MIT administration and Lincoln's other divisions, the solution to the systems engineering problem became apparent to the principals. At a January 18, 1958, meeting in McCormack's Cambridge home, Air Force Secretary James H. Douglas met with MIT's Stratton to work out a solution. Their compromise was to separate Division 6 from the rest of Lincoln as the core of a new nonprofit organization that would handle SAGE and future systems engineering tasks. For its part, the Air Force would strengthen its management of electronic systems to make the job easier. MIT got what it wanted: Lincoln would continue doing research with the substantial independence its personnel desired.[19] The new organization would be called MITRE.

Working with McCormack on the new organization's charter was H. Rowan Gaither, a San Francisco banker who had helped establish the RAND Corporation and was the chairman of RAND's board of trustees. Like RAND and SDC, the new MITRE Corporation would be governed by a board of trustees and have no capital stock. Unlike RAND and SDC, it was limited to working for the government. To appease the blow to SDC, some of the board members also were board members of SDC. The corporation came into official existence in July 1958 with an initial work statement evolved from the original proposal put forth by Robert Everett and John Jacobs. Essentially it described the function of systems engineering in a way very similar to that of the Ramo-Wooldridge Corporation.[20]

Profiting from the Inside: From TRW to Aerospace Corporation

Although Robert Everett and John Jacobs would model MITRE's tasks on the bases of their own experiences and those of R-W, in other ways the Air Force wanted to avoid the R-W model. R-W had proven its technical abilities and utility in Schriever's organization, but in other ways it was nothing but trouble. The

John F. Jacobs (standing) with Robert R. Everett at Lincoln Laboratory shortly before their transfer to MITRE in 1958. At MITRE, Everett served as president, chief executive officer, and chairman of the board, and Jacobs served as senior vice president. Jacobs headed the systems office at Lincoln, one of the first systems engineering organizations.

main problem was that R-W was a for-profit corporation in which the concerns of shareholders for growth and large dividends necessarily influenced events.

From the beginning of the Western Development Division, industry complained bitterly about R-W's "insider" position. Competitive issues drove their complaints. Aircraft industry leaders believed that the ideal systems approach to weapons development was for the Air Force to let prime contracts to a single contractor to integrate a weapon, a position supported by the Air Force's own regulations on the subject. These same regulations stated that in exceptional circumstances the Air Force itself could act as the prime contractor, an exception that Schriever used to run the intercontinental ballistic missiles program. Industry leaders also complained that the Air Force was creating a new, powerful competitor with close ties to Air Force planning and a concomitant edge in bidding.[21]

To protect his organization from criticism, Schriever enforced a hardware ban on R-W to keep it from acquiring lucrative hardware contracts on any programs in which it was involved as technical direction contractor. In addition, R-W walled off from the rest of the company the technical direction work of Space Technology Laboratories. Continuing concerns led R-W to establish a physically separated second location for its headquarters, in Canoga Park, California. These measures, however, did not satisfy industrial leaders who continued to complain and who kept information away from R-W as much as possible in those projects in which it was technical direction contractor.[22]

Despite the hardware ban against R-W for ballistic missile components, the firm aggressively pursued hardware production projects and contracts elsewhere. Both Ramo and Wooldridge believed that their company could not grow without manufacturing capabilities so they grasped every opportunity to break into such areas as specialized computers for process control, semiconductors, and a variety of components for aircraft and air-breathing missiles.

Shortly after the Soviet Union launched Sputnik in October 1957, R-W received permission to build ballistic missile hardware to test ablative nose cones built by General Electric. Work on the newly named Able program led STL to bid on a contract for a lunar probe in the early rush for space firsts. Strongly backed by Col. Charles Terhune, Schriever's technical director, STL built the Able 1 probe, launched in August 1958. That launch failed when the launcher exploded. STL then built the Pioneer 1 spacecraft. It was launched by the National Aeronautics and Space Administration (NASA) in October 1958 and set a new altitude record of eighty thousand miles from earth. These STL activities fomented even more severe protests from industry, which saw the hardware ban against R-W evaporating.[23]

Funding these ventures stressed R-W's finances. Its computer and semiconductor product lines, in particular, demanded large amounts of capital. Ramo and Wooldridge leaned on their original investor, Thompson Products, for cash to expand facilities and capital equipment. The ensuing negotiations led to an agreement that essentially guaranteed a merger of the two companies. Thompson

An Atlas ICBM topped with an ablative nose cone for a reentry test mission at the Air Force Missile Test Center.

Products gave R-W the cash it desired in return for stock that by 1966 would give Thompson Products more than 80 percent control of R-W. With a downturn in its normal electronic and mechanical components sold to the aircraft and automotive industries in the late 1950s, Thompson Products saw R-W as an investment in the new and growing fields of computing, missiles, and space. Negotiations between the two companies led to their merger, effective October 31, 1958, as Thompson-Ramo-Wooldridge, or TRW.

TRW executives, including Ramo and Wooldridge, recognized STL's awkward position in the new company, so they established STL as an independent subsidiary corporation with its own board of directors. The board was chaired by Jimmy Doolittle, a war hero with impeccable credentials and even more impressive ties to the Air Force and NASA. Just prior to his accession to the STL board chairmanship, he had been head of the National Advisory Committee for Aeronautics, the organizational ancestor of NASA. Before that, as we have seen, he was the appointed arbitrator between Air Materiel Command and ARDC upon the creation of ARDC. No TRW board member or senior manager sat on STL's board. TRW executives recognized that they might have to divest STL and through this reorganization were prepared to do so.[24]

Although TRW might have been prepared to divest STL, neither TRW nor Schriever really wanted that to happen. TRW enjoyed a significant profit from STL and Schriever wanted the experience of STL's personnel with the Air Force's ongoing ICBM and space programs. However, STL's increasing involvement with space projects did not mesh with the political need to separate the laboratories from hardware development contracts. Although Explorer 6 and Pioneer 5 were striking technical accomplishments for TRW and NASA's space projects, they fueled industry's complaints about conflicts of interest. Those complaints led to congressional hearings in February and March 1959.

The hearings, chaired by Representative Chet Holifield of California, featured vehement attacks against STL's "intimate and privileged position" with the Air Force and equally strong defenses of it by Schriever and by TRW executives Simon Ramo and Louis Dunn. The situation demanded some other solution. A plan to sell STL to public investors fell through when Air Force Secretary Douglas vetoed it on the grounds that STL would remain a problem as long as private owners used STL to make a profit. The Holifield Committee's final report seconded that idea and urged that STL be converted into a nonprofit corporation like RAND and MITRE.

In June 1960 the formation of the nonprofit Aerospace Corporation was announced by (left to right) Trevor Gardner, Charles Lauritsen, Roswell Gilpatric, Joseph Charyk, and Lt. Gen. Bernard Schriever. Schriever was reluctant to create Aerospace because he wanted to maintain R-W's support for space and missile programs.

The USAF and the Culture of Innovation

Fearing that they would lose STL's most skilled personnel if STL were so converted, Schriever initially opposed conversion. However, a further investigation led by California Institute of Technology professor Clark Millikan at the behest of Secretary Douglas recommended conversion. The investigating committee included a number of personnel who had been involved with the initial selection of Ramo-Wooldridge to perform systems engineering for Atlas. They proposed that four of STL's functions—advanced planning and evaluation, initial system design, technical evaluation of contractor proposals, and contractor technical monitoring—should be placed in a nonprofit corporation. Schriever reluctantly agreed and the Aerospace Corporation was formed on June 4, 1960.

At Schriever's insistence, STL continued its systems engineering and technical direction role for the ballistic missile programs, but all other efforts, including the growing space programs, were transferred to Aerospace. Dr. Ivan Getting left his vice presidency at Raytheon to become Aerospace's first president and a number of STL personnel transferred to the new corporation. This reorganization ended the controversy about TRW's insider position with the Air Force, but as industry had feared there was a powerful new competitor to contend with as a result of the Air Force's policies and TRW's astute and aggressive management.[25]

Ad Hoc Organization for Electronics

While civilian organizations like STL, Lincoln, and MITRE evolved to manage the technical development of ballistic missile, command and control, and space systems, the Air Force had to organize itself internally for those same efforts. Ballistic missiles programs benefited greatly from Schriever's strong leadership and the top priority that he acquired for his programs, despite skepticism among the Air Force's top brass. SAGE and other electronic systems did not enjoy such strong Air Force leadership, and as a result their effort was weak and disorganized.

The Air Force's organization for SAGE began with ARDC's assignment of its AFCRC to work with Lincoln and to integrate Lincoln's technologies into the rest of the Air Force—in this case, ADC. ADC supplied its 4620th Air Defense Wing to work directly with Lincoln, which was working semi-independently from AFCRC. In the meantime, ADC promoted Michigan's system. When Michigan lost its duel with MIT and the Air Force adopted Lincoln's system as the sole source for a new air defense system, the Air Force officially created a joint ARDC–AMC project office to run the program. This was in accordance with its standard Weapon System Project Office procedure in place in 1954. The Air Defense Engineering Services project office in New York then coordinated the efforts of Lincoln, IBM, the 4620th Air Defense Wing, and AFCRC in the installation of SAGE across North America, implemented with Western Electric and Bell Telephone Laboratories. The need for hundreds of programmers to adapt the master program at the various continental SAGE sites allowed ADC to bring RAND into the picture, along with its training and simulation capability. Through RAND and the 4620th Air Defense Wing, ADC soon installed itself into the ADES office in a much stronger manner than officially sanctioned in the Weapon System Project Office regulations.*

Despite the cumbersome interrelationships of AFCRC, ADES, the 4620th Air Defense Wing, ARDC, ADC, and AMC, the SAGE project moved along reasonably well until the problem of integration reared its ugly head. At that point the informal relationships among these organizations became a liability. The aircraft contractors who built the new interceptors and missiles, the Navy with its picket ships, and the Army with its antiaircraft missiles all had their hands full with making their technologies work. They had little incentive to modify their systems or expend substantial efforts to understand SAGE. Nor did their government contract monitors have the authority, the responsibility, or in some cases the desire to move their projects under SAGE's control. ARDC, the one organization that might have had authority for the Air Force, had no real power over Lincoln, and even if it had, ARDC could not command the Army or Navy. Western Electric would not take responsibility after losing its bid for power against Lincoln. For its part, Lincoln's desire to control SAGE informally contradicted MIT's desire to escape this long-term, nonresearch responsibility.

Only a unified, triservice organization really could solve the integration problem, but service jealousies worked against such a solution. Furthermore, the intricate, politically sensitive interactions with MIT complicated the Air Force's efforts to take control of SAGE. After all, the Air Force's senior officers had placed MIT instead of ARDC in control of SAGE development. By 1956, however, MIT's determination to remove itself from SAGE provided the Air Force with the opportunity to take control of its own technologies.

To help with the integration problem, ARDC created an Electronic Supporting Systems Division within its Directorate of Systems Management in March 1956.[26] Brig. Gen. I. L. Farman ran the division, and in November 1956 he held a Continental Air Defense Conference at ARDC Headquarters in Baltimore. Three major problems dominated the discussion: the problem of integration across the Air Force and other services; the uncertainty about Lincoln's attitudes, plans, and roles; and the role of ADC. Frustrated with their lack of control over Lincoln, conferees noted that the laboratory could not be pinned down easily because personnel there were "never willing to put anything in writing." Everyone agreed that ARDC and AMC needed to expand and strengthen their joint project office, but ADC proposed full partnership in the project office instead of taking its customary role by having a liaison. ARDC and AMC united against this threat, squashing for the moment any major reorganization that would call into question the basic ARDC–AMC agreements about project offices to run large-scale

---

*See Chapter 4 for details.

programs. Col. Al Shiely of ARDC and Col. O. M. Scott of AMC were designated the joint project officers and were charged with recommending proposals to strengthen the office.[27]

On December 6 General Farman held a second conference to hear Shiely and Scott's proposals. Surprisingly, Shiely and Scott proposed and senior officers agreed that ADC become a full participant in the project office. They next presented a number of alternatives for the systems engineering contractor, and recommended that Bell Telephone Laboratories be asked to take the job. If Bell refused, they would next request that RAND create a separate organization to perform systems engineering. Only if neither of those alternatives worked would they ask MIT. (As we have seen, BTL refused, the RAND alternative did not work out, and the Air Force eventually asked MIT to deal with the problem.) Shiely and Scott then presented their proposal to the Air Staff on December 14, 1956.[28]

Temporarily establishing the SAGE Weapons Integration Group to handle immediate problems, the Air Force considered the Shiely–Scott proposal. The scope of their recommendations was such that Air Force Chief of Staff Nathan Twining appointed Brig. Gen. Donald Hutchinson to investigate the problem. Hutchinson's investigation ended in April 1957 when he presented his results to the Air Council and then to Secretary of the Air Force Douglas. The Air Council "stated a belief that a field agency similar to the Western Development Division for Ballistic Missiles and that Special Assistants in AMC, ADC, and ARDC are necessary in order to coordinate the technical aspects of the program."[29]

Schriever's organization was a negative example in the mind of ARDC commander Thomas Power. Writing to Twining on April 25, 1957, Power stated that such an organization placed a great strain on his organization. In Power's mind the necessity for secrecy in establishing the ICBM effort was paramount and was not relevant for air defense systems. He asserted that "4 or 5 WDD's all operating at the same time cannot possibly be supported to the extent necessary to make each a successful operation"[30] because the special assistants required to work with Schriever's organization gave orders to Power's staff just as Power did, thus making his staff responsible to two commanders. Power then threatened ultimate upheaval:

> going down this road of special organizations and Special Assistant Commanders leads to the interesting thought that if you continue in this direction you are, in effect, vitiating the Command Headquarters. . . . It probably would be more effective, if this trend continues, to do away with this Command and operate directly from Washington. This would not save resources and time and would ultimately result in a very inefficient operation and finally in chaos.[31]

Lt. Gen. Thomas Power (center), then commander of ARDC, inspected the Atlas missile production line at the Convair Division plant in San Diego, California, in March 1957. To Power's right is Convair manager James Dempsey. Dempsey was a significant force in the battle between Convair and Schriever's organization over the role of Ramo-Wooldridge.

Power's argument carried the day and Twining directed the creation of the new Air Defense Systems Management Office (ADSMO) in June 1957. Established at Hanscom AFB next to Lincoln Laboratory and AFCRC, ADSMO did not have the directive authority of Schriever's Western Development Division. In a briefing to the commanders of ARDC, AMC, and ADC, Col. Charles Thorpe, the chief of the new office, reassured the commanders:

> ADSMO does not intend to replace or duplicate the effort now being performed by various agencies. We may recommend combining some of the effort, if it is not properly directed. We are a joint management agency responsible to the three major commands involved in air defense. *We do not direct*—we dot [*sic*] not replace or assume any command prerogatives. We would do this only as you would desire us to assume this responsibility.[32] [Thorpe's italics]

The new office would bring in more personnel to coordinate activities and would work directly with new positions in the Air Staff to handle coordination with the Army and Navy. ADSMO coordinated not only SAGE but also the development of the Distant Early Warning (DEW) radar line along the Arctic Ocean, and other air defense improvements and projects. The ADES office for SAGE became the 216L project office, one of several so-called L systems that managed electronic systems. It became one of several Electronic Supporting System Project Offices (ESSPOs) that the Air Force was beginning to realize formed a class of technical projects separate from its typical aircraft and missile projects.[33]

Unfortunately, ADSMO's weaknesses made its job very difficult. Because the systems engineering contractor situation had not been resolved, it had little systems engineering capability. It had no authority over ADES or the other ESSPOs. Its commander was only a colonel, which signaled to everyone that the office was not very important. Finally, it could make only recommendations that had to be coordinated with all three major commands.

These problems soon became apparent and ARDC's next move was to strengthen ADSMO by assigning a general officer to command it. In February 1958 Maj. Gen. Kenneth Bergquist became the first commander of the office, now named the Air Defense Systems Integration Division (ADSID). ARDC also named Bergquist the deputy commander for systems integration, which ensured his authority in ARDC. ARDC did not, however, change the office's status with AMC or ADC although the Air Force issued a new regulation describing the office's functions and duties. The first draft of Regulation 20-13 placed Bergquist in a similar position and with similar authority to Schriever and his WDD. However, the Air Staff soon watered that down to leave ADSID's new commander with little more authority than Colonel Thorpe had possessed.[34]

The formation of ADSID, which corresponded in time with negotiations regarding the creation of MITRE, was theoretically to resolve the problem of integration. Although nominally under ARDC, Bergquist understood his job to require a broader vision and he worked directly with the Air Staff. Because integration also required substantial interactions with the Army and Navy, Bergquist came to believe that the solution to the air defense problem required that ADSID work as an office of the secretary of defense, above all three services. Having been in Hawaii in air defense at the time of the Japanese attack on

Pearl Harbor, Bergquist believed strongly that solving the air defense problem was crucial. Apparently so did Air Force Secretary Douglas for Bergquist later recalled discussions about placing ADSID in the Department of Defense and having MITRE work to integrate the services at that level.[35]

Bergquist persisted in that idea through late 1958 and early 1959. He and the ADSID staff worked closely with the Directorate of Research and Engineering and the Weapon System Evaluation Group in the Office of the Secretary of Defense, with North American Air Defense Command (NORAD), and the Royal Canadian Air Force. Because moving to the OSD was not in the cards, Bergquist suggested instead that ADSID be transferred to NORAD. The Air Staff denied his recommendation in April 1959 because ADSID seemed to be working well in its present configuration.[36]

MITRE also hoped to continue its work on air defense integration at the highest level just as it had done as part of Lincoln Laboratory. In the negotiations that led to MITRE's creation, however, MIT's hurry to escape from systems engineering weakened MITRE's position, and in the final contract MITRE did not have the kind of directive authority that R-W had enjoyed. Rather, MITRE was to act as a technical consultant under guidance from ADSID. ADSID would function as the systems engineer and MITRE would assist with that task. To the extent that ADSID worked at the highest levels to integrate air defenses, so, too, would MITRE. Unfortunately for both entities, that was not in the cards. Instead of moving up the chain to control all air defenses, ADSID would be abolished and its successor moved down the chain toward implementation of specific technologies.[37]

Schriever Takes Command

From late 1956 until the creation of ADSID in 1958, General Bergquist and others who seriously desired strong air defenses looked longingly at the authority and effectiveness of Schriever's WDD and the "western complex" in Inglewood. Efforts to give air defense the same authority as Schriever's group, however, ran into determined opposition from ARDC and AMC. Those commands objected on the grounds of disruption to their command structures and the power usurped by Schriever's group. It was difficult enough to serve Schriever alone; the demands of several such groups would wreak havoc. Ironically, the events that finally terminated ADSID and its efforts for independence eventually involved Schriever himself when he took command of ARDC in 1959.

The restructuring of the Air Force's R&D organizations can be traced to soul-searching in the wake of the Soviet launch of Sputnik in October 1957. Shortly after that launch, Air Force Chief of Staff Thomas White asked the Scientific Advisory Board to "conduct an impartial and searching review of the organization, functions, policies and procedures of the Air Force and Air Research and Development Command in relation to accomplishments in research and development over the past seven years." SAB chairman James Doolittle appointed a committee headed by MIT physics professor H. Guyford Stever to respond to White's request.[38]

The Stever Committee Report appeared in June 1958 and had a number of important recommendations. First, and perhaps most important, the committee recommended that the Air Force's tendency to centralize weapon systems authority at ARDC Headquarters and with the Air Staff needed to be reversed. Harkening back to the problems that existed when ARDC was created in 1950, ARDC needed to delegate project authority to the working officers and engineers on the projects. This would include authority over the project's funding. The committee also recommended that ARDC be reorganized along functional lines, meaning that deputy commanders should head efforts in research, technical development, weapon systems, and testing, and be given complete control over the programs in those phases. The committee also noted that AMC controlled 80 percent of the funding and prepared the R&D budget recommendations to higher authorities. Instead, ARDC should perform this task and control their own R&D funding. Budgets should be prepared along program lines, similar to the Gillette Procedures used for Schriever's ballistic missile programs.[39]

Gen. Samuel Anderson, then head of ARDC, followed up the Stever Committee Report with an internal study, eventually known as the Anderson Committee Report. That report, issued in February 1959, concurred with the Stever recommendations except for the functional reorganization. Instead, the committee recommended that ARDC's functions be realigned into four major "product" groupings: aerodynamics, ballistic and space systems, electronic systems, and basic research. This idea for organization structure would stick. Because Anderson was to leave his post to become head of AMC, he decided to leave the committee recommendations for the next commander, Lt. Gen. Bernard Schriever.[40]

The Air Staff wanted Schriever to fix the continuing organizational problems with the electronic systems. In April 1959 Schriever became the head of ARDC, and in May he launched his own study, headed by Col. Jewell Maxwell, to focus explicitly on the organization of ARDC, incorporating the principle of concurrency. The Maxwell Group Report, submitted at the end of July, agreed with the organization structure recommended by the Anderson group, and specified the duties of ARDC Headquarters with respect to the "Western, Central, and Eastern" systems divisions. Headquarters would focus on long-range planning, the relationships among the programs, intelligence, and "final program review."[41]

At the same time that the Maxwell group was working, the Air Staff had appointed a high-level group headed by AMC's commander, General Anderson, to examine the entire Air Force procedure for managing large systems. In May 1959 Air Force Vice Chief of Staff Curtis LeMay established the "Weapons Systems Study Group" headed by Anderson to investigate the applicability of ballistic missile "concurrency" methods throughout the Air Force. This group comprised a number of senior officers, including General Schriever. Both the Maxwell group for ARDC and the Anderson committee at the Air Staff level worked under the assumption that the Air Force would reorganize based on the three major systems areas—ballistic missiles and space, aerodynamic systems, and electronics. To ensure compatibility between ARDC and AMC efforts, the two commands established an ad hoc committee to foster communications and coordinate their planning.[42]

ARDC and AMC already had joint offices for ballistic missiles in Inglewood and for aircraft and other aerodynamic systems in Dayton, Ohio. Neither command, however, had equivalent divisions for electronic systems. If ARDC established an electronics division, then, in keeping with the Cook–Craigie Procedures, AMC would need to establish its sister center alongside. Because Schriever was moving forward with the ARDC reorganization, AMC stepped up its efforts.

Schriever formed another group of experts in August and September to look at his plans for the "Eastern Development Force." The group, which included Guyford Stever and Jerome Wiesner from MIT, William Baker from Bell Laboratories, Rube Mettler from Space Technology Laboratories, and Ivan Getting from Raytheon, gave Schriever their blessing. In September, with the concurrence of the Air Staff, Schriever implemented the Maxwell Report recommendations. Effective November 16, ARDC established the Air Force Command and Control Development Division (AFCCDD), and Anderson's AMC established its Electronics Systems Center (ESC) at Hanscom Field, effective January 1, 1960.[43]

The formation of what became known as the Hanscom Complex involved much more than establishing ARDC and AMC offices there. For one thing, it required the transfer of eleven ESSPOs from Dayton and New York to Hanscom. Second, it reignited the old fight between AMC's Rome Air Development Center and Hanscom as to the final location of the complex. The existence of the MITRE Corporation near Hanscom was one of the crucial factors leading to the Hanscom decision. Third, there was the vexing problem of air defense integration and what to do about Bergquist's ADSID.[44]

Although everyone agreed that the integration problem involved the entire air defense establishments of the Army, Navy, and Air Force, along with the Canadian Air Force, Bergquist's efforts to move ADSID (and MITRE) to the DOD or to NORAD had failed. With the reorganization coming, Bergquist pressed his case with Schriever, but by Thanksgiving he realized that the fight to save ADSID was futile. He and Schriever soon met to work out the new arrangements, which folded ADSID's planning and coordinating functions into AFCCDD and place Bergquist in command of the new division.[45]

In early 1960 AFCCDD staff negotiated with ARDC the functions of their organization and proposed its structure and processes. Although most of these were mundane, the Inglewood model advanced by ARDC clashed with Hanscom's experts on the issues of advanced planning and integration. Fundamental to the Inglewood model was the notion that experts could determine with some accuracy the functions and requirements of the weapon system at the start of a program. The function of the ARDC division (Western Development Division, for example) was then to translate these firm requirements into a system, including the vehicle, ground systems, and training. The command that ultimately would use the system, such as SAC, played a very small role at the start.

By contrast, the experience of Lincoln Laboratory and MITRE Corporation personnel, along with those at AFCRC, the ADES office, and ADC suggested that firm requirements were unlikely at the start and the understanding of the system would evolve over time. Because of the close ties between air defense tactics and the functions to be programmed into the computer, the end-user command had to be involved in a significant way right at the start. Furthermore, understanding how best to operate the system and what functions to program into the computer evolved over time as the operators became familiar with the system, and the computer developers and programmers became familiar with operations. Finally, integrating new sensors and weapons caused major changes because any of these new technologies could change the operating parameters and the organization of air defense significantly. Those changes in turn required modifying the computer program, a costly and time-consuming process.[46]

Conflict over these contrasting "philosophies" of R&D appeared when AFC-CDD and MITRE presented their ideas for organization structure and procedures to ARDC Headquarters for approval. The AFCCDD proposal stated that all other Air Force commands would be responsible for ensuring the standardization among systems necessary for integration into AFCCDD's command and control systems. It also proposed the creation of a Directorate of Advanced Studies, which would perform the functions of ADSID, namely the analysis of future systems and the integration of such systems with various weapons and technologies. ARDC objected to the proviso for other commands to ensure standardization, and the ADSID functions ultimately were diffused into a number of AFCCDD organizations. Schriever appointed Brig. Gen. Charles Terhune, his technical director from Inglewood, as the deputy commander of AFCCDD to help ensure that the division moved in the right direction. Consequently, AFCCDD wound up with an organization very similar to the Inglewood model.[47]

As those debates began in late 1959, ARDC commander Schriever requested a comprehensive study of the Air Force's electronic, or L systems. He wanted an assessment of the "technical realism" of these systems by a mixed group of scientists, engineers, and officers from industry, academia, and the military. Schriever hoped to use the results of this study to inform and validate his decisions about the organization of AFCCDD and about the management of L systems in general. John Jacobs of MITRE organized the study which began in late November. The first technical panel meetings began in January and continued past the original April deadline through June 1960. This group, which produced the so-called Winter Study, soon took a direction unforeseen by Schriever and ARDC Headquarters.[48]

Instead of confirming Schriever's Inglewood model of rapid development based on concurrency, the Winter Study Report was a ringing endorsement of the Hanscom–Lincoln–MITRE viewpoint about command and control systems. Right from the start the report emphasized the connection between military doctrine, strategy, tactics, and the specific technical characteristics of C^2 systems. For example, a nuclear first-strike strategy places emphasis on control and sensor systems, but a second, retaliatory strike requires survival of the force and a basic C^2 function in hardened and redundant facilities. Instead of focusing on the gritty technical details of each system, the group indicted the military's indecision regarding strategy and tactics.

Furthermore, to develop an appropriate C^2 system, functional integration was far more important than technical integration. That is, the military needed to integrate its organizations, along with appropriate strategy, tactics, and procedures, before a C^2 system could automate those functions. Even when functional integration occurred, the engineering groups should work closely on operational experiments with the end-user commands before committing to hardware development and production. To make matters worse, the group recommended concentrated "responsibility and authority for making command and control system engineering decisions. . . ," most appropriately at the DOD level. Current L systems used different assumptions, tactics, and procedures, symptoms of the underlying incompatibilities and lack of coordination among the military services. Although improving technologies would be helpful, the main problem was that "our present capabilities to build hardware components for use in systems is well beyond our knowledge of how to assemble these components into efficient systems." The main problem was social organization, not technical capability.[49]

The results of the study were not surprising to old hands in air defense at AFCCDD, ESC, or MITRE, but were anathema to ARDC Headquarters. They bucked the ARDC emphasis on up-front planning followed by rapid development using concurrency, and made recommendations far beyond ARDC's authority. ARDC demanded the creation of a "sanitized" version of the report, except for the full copies of the original distributed within Hanscom. In effect, the philosophy propounded by the Winter Study Group might suit the specific problems of C^2 systems but would confound and confuse those outside of this environment. The sanitized version finally appeared on March 31, 1961, almost nine months after the original report appeared.[50]

In the meantime, General Bergquist and the AFCCDD moved forward with Winter Study Group recommendations as best they could, given the limits to their authority. To eliminate duplication and set all of the L systems on a consistent basis, AFCCDD created its own conceptual framework regarding strategy, tactics, and operations. Consistent with prior practices, they also worked to obtain operator participation through design and development. Because "neither the environment nor the operational concept of the user are static or stable elements, it has been necessary to continuously review and revise them and to re-evaluate each system design in relation to these changes." Finally, they began to expand their in-house study capability, which eventually led to creation of a Directorate of Advanced Planning in December 1960. ADSID was dissolved in October 1960 but the new directorate essentially recreated its functionality to plan for and coordinate L systems and to integrate new weapons and technologies. With this directorate AFCCDD created the organizational structures and processes necessary to complete its mission, modifying the Inglewood model sufficiently to succeed.[51]

MITRE, too, had to adjust to the arrival of Schriever's style of management. MITRE management expected that its contingent of former Lincoln Division 6 personnel would continue to work on the problem of large-scale integration and systems engineering for SAGE and the various new C^2 systems under development for the Air Force—systems for SAC, ballistic missile early warning, NORAD, satellite tracking, air traffic control, and so on. With the creation of AFCCDD and the impending dissolution of ADSID, however, MITRE's role seemed likely to change.

One key issue at stake was MITRE's assistance to such other government agencies as the Federal Aviation Administration (FAA) or the OSD. MITRE already was working under contract for the FAA on a possible new air traffic control system that worked with both military and civilian inputs. The more important issue was the corporation's relationship to the OSD. The Air Force preferred to have a single contract with MITRE, and to have any other government agencies transfer their funds to the Air Force before funneling it into the MITRE contract. It also did not mind if MITRE provided assistance to the OSD for that might be a mechanism by which to influence OSD decisions. MITRE, on the other hand, preferred separate contracts to maintain a semblance of independence. To remind the Air Force of Air Force Secretary Douglas's oral statements that MITRE should work with the OSD to assist with the integration issue, members of MITRE's board and management visited Washington, D.C., in January 1960. The results of that meeting seemed optimistic at first, but in the end words meant little. The OSD did not provide a contract and MITRE remained thoroughly dependent on the Air Force.

The basic issue at stake was whether MITRE would work independently to influence military doctrine and technologies inside the U.S. military as a whole, as the "principal adviser" that Lincoln Laboratory had been, or would be reoriented along the lines of Ramo-Wooldridge so that it acted as a systems engineering contractor to the Air Force to develop specific projects. MITRE leadership wanted to maintain independence and influence to develop new command and control systems for the military, but the object of Schriever's reforms was to make MITRE a tool of the Air Force for specific projects.

In the end, the Air Force controlled the funding, and by withholding funds except for specific projects in the fall of 1960 it forced MITRE to do its bidding. MITRE reorganized its structure to match that of AFCCDD just as Ramo-Wooldridge had done with Schriever's WDD. The Air Force withheld funds again

In this early automated en route air traffic control laboratory, MITRE developed methods to show how SAGE facilities could simultaneously perform air traffic control for Federal Aviation Administration civil systems and air defense functions.

in 1962 to force through a significant revision to the MITRE contract to make the corporation more "responsive." The new contract replaced the term "principal adviser" with a description of MITRE's duties as "research and development for system design, system engineering, technical direction, intersystem integration, and research and experimentation to achieve continuing advances in the complete field of Command and Control Systems." To quote Deputy Commander Charles Terhune, there needed to be a "marriage" of MITRE and the AFC-CDD to accomplish program objectives. In effect, MITRE became the East Coast electronics version of the Aerospace Corporation. The Inglewood model's transfer to the East Coast was complete.[52]

Standardizing Systems Management

While Schriever's reform of ARDC reorganized and reoriented the Air Force's and MITRE's efforts on the East Coast, ongoing deliberations regarding the applicability of the Inglewood model across the entire Air Force were under way in the Weapons Systems Study Group. This group, also known as the

Anderson Committee, included Anderson, Schriever, and a number of other senior officers. The committee agreed that the methods used in Inglewood should be adopted, with the planning and implementation of new systems on a systems, or "life cycle," basis. Planning for the entire system would occur up front and the project offices would have the authority to manage the development, including authority over the funding. However, the committee split into three camps regarding the actual organization.

In June 1960 the Air Staff reviewed the group's recommendations, which essentially boiled down to a presentation of the three plans. The least ambitious plan, sponsored by Maj. Gen. Mark Bradley, the acting DCS/M, kept the Cook–Craigie arrangement of WSPOs and added a more significant degree of input from the end-user commands. ARDC and AMC would keep their existing roles but extend the Inglewood methods to aeronautics and electronics. Anderson, AMC commander, recommended recombining ARDC and AMC. Schriever, ARDC commander, proposed a new split of functions that would give ARDC the entire acquisition process and leave long-term logistics to the remnants of AMC. The Air Staff, led by Gen. Thomas White, selected the Bradley proposal and ordered the respective commands to implement its facets. That implementation included installing a new set of procedures for the entire Air Force modeled on Schriever's organizational processes at Inglewood.[53]

Those new processes, ultimately known as the "375-series" of regulations for systems management, originated with one of Schriever's officers in the Ballistic Missile Division, Col. Ben Bellis, who thought that they ought to document the procedures developed at the Inglewood complex and who worked with others at BMD to write them down. Apparently during the deliberations of the Anderson Committee, the existence of this documentation came to light. The one point on which all committee members agreed was that after appropriate review those processes should be implemented for all major systems, including ballistic missiles and space, aeronautics, and electronics. Because the selection of the Bradley proposal essentially stated that the Inglewood processes and organization structure should be copied, Bellis' procedures went through a formal review. The first three regulations were published on August 31, 1960, and subsequently were revised and extended.[54]

Air Force Regulation (AFR) 375-1, *Systems Management, Management of System Programs* defined system programs and systems management, and separated the development process into conception, acquisition, and operations. Systems Management applied primarily to acquisition.[55] AFR 375-2 specified the responsibilities of the System Program Office.[56] AFR 375-3 established the responsibilities of the system program director to whom it gave wide latitude and substantial authority.[57] AFR 375-4 specified the documentation required for the System Program, including the System Package Program, the key program document. Each System Package Program had to provide information in a number of areas, including cost, schedule, management, logistics, operations, training, and

A meeting in the program control and status room in the Air Force Ballistic Missile Division offices, December 1957. This type of room became a standard in aerospace. Developed by Schriever's group, this one may have been the first and helped lend an aura of competence to Schriever's management methods.

security.[58] The final set of regulations dealt with reliability.[59] Air Force officers eventually realized the applicability of the reliability concept to weapons of all kinds and levied the reliability regulations on the entire Air Force.

The 375 regulations formally applied the ARDC–AMC project office concept across the Air Force's other large acquisition programs. They gave project officers substantial authority to carry out their programs without micromanagement from Air Force or ARDC Headquarters, and specified the documents and information required by both headquarters for program approval and periodic reporting. Through their application across ballistic missiles, aeronautics, and electronics, the regulations provided a uniform standard for the commanders of ARDC and AMC to control all of their large projects and, through the methods of concurrency implied therein, to speed them up. Schriever's processes were triumphant, and with the appointment of Robert McNamara to be secretary of defense so, too, would be Schriever's proposed organization for the Air Force.

The issue that catalyzed the Air Force to take drastic action was the growing importance of space technologies and missions. Soon the military services were competing and bickering over space project control, a new "roles and missions" controversy identical in nature to the turf battles over ballistic missiles in the 1940s and 1950s. In the aftermath of Sputnik, the continuing squabbles led the

Eisenhower administration and Congress to create NASA to run the civilian space program, and the Advanced Research Projects Agency (ARPA) to manage the DOD's space programs. The Air Force, which had been running the WS-117L program to develop military reconnaissance satellites, thus found its authority for the project usurped by the new agency. To regain what it believed to be the "natural" extension of its service from air to space, the Air Force would have to prove its managerial, organizational, and technical competence. Fortunately for the Air Force, the ARPA effort foundered and the services once again regained a measure of control over their space projects. The Air Force, however, did not gain the primacy and control of military space that it desired. The next major opportunity for change would come in the new Kennedy administration in 1961.

For Robert McNamara, managerial and organizational factors counted most. Deputy Secretary of Defense Roswell Gilpatric hinted that DOD might assign

Commander of the Air Force Systems Command Gen. Bernard A. Schriever with MITRE president C. W. Halligan. Schriever insisted that MITRE serve as adviser and assistant to the Air Force for electronic systems, just as Aerospace Corporation did for ballistic missiles.

the military space mission to the Air Force on proper reorganization of space efforts into a single command. Schriever had already investigated this possibility in 1960 when he asked Trevor Gardner, then president of Hycon Manufacturing, to head a committee to study the Air Force's organization for space. Gardner's committee recommended that the Air Force separate its space programs into a new division, apart from ballistic missiles. With Gilpatric's carrot in hand, Schriever combined the Gardner Committee recommendations with the "Schriever plan" presented in the previous year to the Anderson Committee. With the Air Force's space mission at stake versus the Army and Navy, Air Force Chief of Staff White approved the previously rejected plan. Secretary of the Air Force Eugene Zuckert and Secretary of Defense McNamara quickly approved the plan, and on March 8, 1961, McMamara transferred all space research to the Air Force.[60]

On March 14, 1961, General White informed all affected commands of the adoption of the Schriever plan, which reallocated the procurement activities of Air Materiel Command to an expanded ARDC, now redesignated Air Force Systems Command (AFSC). AMC itself was deactivated and replaced by a smaller Air Force Logistics Command (AFLC). In addition, the new organization separated research into a distinct organization, the Office of Aerospace Research, reporting directly to Air Force Headquarters. AFSC, which came into being on April 1, 1961, had four major commands: the Ballistic Systems Division (BSD) based in San Bernardino, the Space Systems Division (SSD) in El Segundo, the Aeronautical Systems Division (ASD) in Dayton, and the Electronics Systems Division (ESD) in Lexington.[61]

The adoption of the 375 Procedures and the creation of AFSC with its quartet of systems divisions essentially standardized the Air Force acquisition process on the Inglewood model. However, there were some significant differences. Whereas at Inglewood Schriever headed a joint program office with ARDC and AMC, AFSC eliminated the problems of joint command by combining the functions of both commands. Another difference was that ARDC and AMC were essentially excluded from the decision-making processes at Inglewood through the Gillette Procedures. Schriever, now ARDC/AFSC commander, was willing to place ballistic missiles under ARDC, that is, under *his* command. The 375 Procedures spread systems management throughout the Air Force, but also brought Inglewood back under Air Force control.

Standardization of R&D in AFSC went beyond just the 375 regulations. By defining standard reporting practices managers could identify anomalous behavior more readily. In some cases, technology managers went so far as to define the structure of presentations. For example, Maj. Gen. Osmond Ritland, the head of SSD, established monthly meetings with all program and technology managers in his division. With Aerospace Corporation president Ivan Getting in attendance, Ritland reviewed each project in turn:

> Each program director has 15 to 25 minutes to cover his area. . . . The key to rapid comprehension of sticky management problems is standardization of presentation. Each director of a major program is required to adhere to a set formula so the commander and the staff will know exactly what to expect sequentially and how to interpret what each program or system director is reporting on. In his presentation, the director first covers his status on directed action items which resulted from the last meetings. Next he covers his problem areas in their broadest sense. The details of these problems come later. Then, in turn, he reviews his (1) Government financial charts, (2) Contractor financial charts, (3) Contractor manpower status, (4) Contract change negotiations, (5) Letter contract status, (6) His manpower, both military and technical staff of the Aerospace Corporation, (7) What he is doing to meet certain procurement management objectives. The program director may conclude his presentation with a summary chart if he chooses—all others are standard.[62]

Similarly, Schriever established procedures throughout his organization to standardize communications up and down the line. He went to great lengths to ensure that rapid communication moved in well-defined ways. By mid-1961 Schriever's organization molded status reporting into a highly sophisticated system known as Rainbow Reporting because it presented each element of the system on differently colored pages in a small, brightly packaged booklet that specified the formats and procedures for status reports and presentations. Over the next few years, the procedures evolved to include full program, yearly, and monthly milestone schedules, government and contractor financial data, contractor manpower data, reliability data, procurement data, engineering qualification data, and the so-called special PRESTO Procedures for problems that needed immediate attention. For each item, the procedures gave sample forms along with specifications of the information to be included, and defined when and to whom the information was to be reported. They also specified acceptable formats and technologies to be used for presentations to ensure commonality, enabling the top-level managers to judge the programs on common bases.[63]

Configuration management was another missile innovation ordered into other AFSC commands to tighten project control. By late 1961 the ideas of configuration management spread from missiles to the Hanscom complex, with the formation of "configuration control groups" on L systems, such as the NORAD Combat Operations Center (COC) (COC-425L) project and space object tracking (SPADATS–496L). By 1962 AFSC promoted configuration management across all of its divisions through issuance of its configuration management regulations, AFSCM (Air Force Systems Command Manual) 375-1. In August 1962, for example, the NORAD COC project issued its configuration management instructions, which were similar to the general instructions for configuration management with the exception of software. MITRE and ESD had to create "equivalents" in software for the hardware baselines and end-item specifications and product deliveries. That requirement was not controversial within MITRE or ESD because they essentially had been using configuration management in all but name since the SAGE software crisis in 1955–57.[64]

With the establishment of AFSC and its 375-series regulations, systems management came to have a very specific meaning within the Air Force. It was the application of Inglewood-based, 375-series procedures, including configuration management, to large-scale programs managed by AFSC. These regulations defined the roles of the project manager, the communication and control procedures, key documents, and the reliability procedures for the acquisition (development) phase of the system. By April 1961 Schriever's authority and influence reached its apex as he presided over all major development programs in the Air Force using standardized methods of his own making. His methods shortly would become models for technology management throughout the DOD and beyond. Unfortunately, Schriever soon found that in the hands of Robert McNamara his methods would be modified and diffused in ways that he believed distorted their original intent.[65]

McNamara, Phased Planning, and Central Control

The technical problems of missiles and contractual relationships between the Air Force and industry reinforced the adoption of conservative management and engineering practices, and the growth in power of the federal executive branch drove a similar centralizing trend at the highest levels of the government. From the creation of the OSD in 1947 until the height of McNamara's tenure in the mid-1960s, the OSD grew in power as it pulled critical decisions up the hierarchy and subordinated service interests and rivalries.

The National Security Act of 1947 established the position of secretary of defense, the National Security Council, the Central Intelligence Agency, the Air Force, and the Research and Development Board. The secretary of defense's authority was quite limited, particularly because the services could present their budget requests directly to Congress and administer their services independently of each other and the secretary of defense without fear of being overruled. Similarly, the Research and Development Board could only coordinate activities, without authority to enforce decisions.[66]

Continued bickering among the services made it obvious to those outside the military and even to some inside that the secretary of defense needed more authority. In August 1949 President Harry S Truman signed amendments that strengthened the hand of the secretary of defense. They removed the service secretaries from the National Security Council, eliminated the services' right to appeal directly to Congress or the president, increased the Joint Staff in the DOD, and gave the secretary of defense some authority over the budget. The 1949 act required establishment of a unified budget structure for the DOD and all of the services but denied the secretary the right to starve a particular service of money and thus implicitly change its combat functions.[67]

President Dwight D. Eisenhower further strengthened the role of the secretary of defense. In Reorganization Plan #6, approved by Congress in 1953, the Eisenhower administration increased the secretary's staff to monitor the services and abolished coordinating boards like the Research and Development Board. Instead of being "separately administered" as described in the National Security Act of 1947, the service secretaries were to act as the secretary of defense's "operating managers" and "principal advisers." Eisenhower made clear his desire to "manage" the DOD by appointing businessmen Charles Wilson and Neil McElroy to be the secretaries of defense between 1953 and 1959.[68]

With the services continuing to battle for the development of new weapons, particularly missiles and nuclear weapons, Eisenhower pushed for further changes. The Defense Department Reorganization Act of 1958 created unified operational commands, further enlarged the DOD, and gave more authority to the secretary of defense. To end the bickering over radical new weapons and missions, the act allowed the secretary of defense to reassign combat functions and the development and operation of new weapons without congressional approval, and gave the secretary the authority to create a single supply agency for all of the services. Finally, recognizing the importance of new weapons and technologies, the act created the new post of the director of defense research and engineering, with authority over all research and engineering within the DOD. This sweeping reorganization provided the authority and justification for a strong manager to take control of the department. Such a manager arrived with the next administration.

President John F. Kennedy selected Robert McNamara to be his secretary of defense early in 1961. McNamara trained at the University of California, Berkeley, and taught business courses for a short time at Harvard before World War II. During the war he performed statistical analyses for Army logistics, determining the quantities of replacement parts needed based on statistical assessments of combat and operations. After the war he joined Ford Motor Company where he was tagged as one of the mathematically trained "whiz kids" who reformed Ford's disorganized finances and helped turn the company around. He rose quickly, eventually becoming company president.[69]

McNamara was famous for his belief in centralized control implemented through quantitative measurement. With the authority granted by the Defense Department Reorganization Act of 1958 to withhold funding from the services and transfer assignments among them, he could exert his power effectively. He spent the spring of 1961 gathering information about the department, initiating

Secretary of Defense Robert McNamara (center, foreground) accompanied President John Kennedy (partially obscured, background) to Vandenberg Air Force Base in March 1962.

over one hundred studies known as "McNamara's 100 trombones," or "the 92 Labors of Secretary McNamara."†[70]

Without waiting for completion of the studies, he also installed RAND chief economist Charles Hitch as the DOD comptroller. In early 1961 McNamara listened to Hitch's proposals for financial reform of the strategic nuclear forces. At the end of his presentation McNamara exclaimed, "That's exactly what I want [but with] . . . one change. Do it for the entire defense program. And in less than a year." This decision and its implementation occurred so quickly that when the services realized what was happening, it was a fait accompli.[71]

The essence of Hitch's Program Planning and Budgeting System (PPBS) was the allocation of funding by programs or "missions." Instead of funding military activities through the "line" organizations—sending it to the Army, Navy,

---

†McNamara initially instituted ninety-two projects but these grew to over one hundred, hence the number differences.

or Air Force—the DOD would fund military programs such as Strategic Offensive and Defensive Forces, Central Supply and Maintenance, or R&D, regardless of the service. Each of those areas then would project five years into the future, coordinate with each service, and send the package to the secretary of defense. This new process became his primary means of controlling the military because through it he controlled how much money each service spent and where it spent the money.

Given McNamara's background as a financial manager at Ford, and Hitch's qualifications as an economist, it was not surprising that they considered economic criteria to be foremost in making decisions for future weapon systems.[72] Hitch's PPBS required that life-cycle cost estimates be performed before deciding whether to develop a new weapon system. This agreed with the result of one of McNamara's studies: "Shortening Development Time and Reducing Development and Systems Cost." The study claimed that "reducing lead time and cost" should be made equal in priority to performance. It deemphasized pushing the state of the art and required that "feasibility and effectiveness studies" calculate technical risks and cost-to-effectiveness ratios.[73]

Following up on that study, McNamara assigned the task of improving R&D management to John Rubel, the deputy director of defense research and engineering, in September 1961. Rubel proposed to set up management model programs whose methods could then be copied throughout all of the services, starting with the Air Force Agena, TFX fighter, Titan III, and Medium Range Ballistic Missile programs.[74] Rubel proposed a Phase I effort to develop a preliminary design. That effort would ensure "that the cost estimates for the subsequent development effort are based on a solid foundation." The preliminary design effort would generate "a set of drawings and specifications and descriptive documents" to describe methods for managing the program, including schedules, milestones, tasks, objectives, and policies.[75] He stated, "Strong centralized project-type organization must be insisted upon for all major elements. . . ."[76] Rubel had no reservations about forcing industrial contractors to organize and manage their projects in the way he wanted; if they wanted the job, they had to conform.

He made clear in the requests for proposals that a go-ahead for Phase I *did not* constitute program approval. Previously, award of a preliminary design contract constituted de facto project approval for development and production. Now, however, only the secretary of defense could approve a project, and he would not do so until Phase I was completed and the program reviewed.[77] Rubel stated that "the fact that improved definition is required before larger-scale commitments are undertaken is neither surprising nor unique, although it is true that on most programs this definition phase has been less clearly identifiable since it has been stretched out in time and interwoven with other program activities such as development, model fabrication, testing and, in some cases, even production." Rubel did not believe that a program definition phase would slow high-priority programs. "In fact," he wrote, "our real progress should be accelerated as the result of obtaining a better focusing of our efforts."[78]

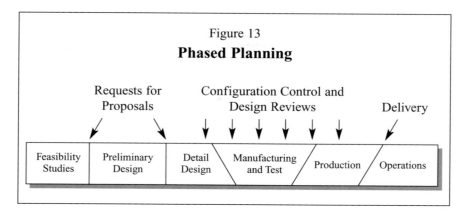

Figure 13

Phased Planning

The phased approach (see Figure 13) brought several benefits to upper management. It ensured better cost, schedule, and technical definition. If the contractor or agency did not provide appropriate information, management could cancel or modify the program. Organizations therefore made strenuous efforts to finalize a design and estimate its cost. The preliminary design phase provided management with a decision point before spending large sums of money, making projects easier to terminate and contractors easier to control. From the Air Force, phased planning spread throughout the DOD.

Rubel wrote a general discussion paper, known in the services as "The Rubel Philosophy," which outlined problems and approaches to management in development programs. His objectives—reduction of lead times and costs, better use of engineering and scientific talent, improved quality, and enhanced coordination with NASA—were not controversial. His proposed solutions, however, spurred debate: standardized reporting, reducing the use of cost reimbursement contracts, improving information in contract bids and evaluations, and collecting personnel and cost statistics. Rubel's memo became a lightning rod for debate within the services from the moment of its distribution in October 1961.[79]

Air Force Col. Jewell Maxwell, assistant for management policy, commented on Rubel's policies:

> ... definition and development implies a tendency toward the "fly-before-you-buy" concept which will obviate to some extent application of the concurrency principle during the earlier phases of urgent system acquisition programs.... It has been conservatively estimated that an additional 6 to 9 months will be required to complete the MMRBM development program under the principles established in the DDR&E [Defense Department Research and Engineering, Rubel] memorandum... if too much emphasis is placed on achieving finality of definition

during that phase of the program when data are most meager and decisions are most subjective.[80]

Maxwell saw the earlier emphasis on rapid system deployment being modified to reduce cost, but a more serious problem was a tendency toward "creeping centralization" at the secretary of defense level. Maxwell and other officers perceived a "trend toward imposition of super-management organization at the top of current review and approval channels." The problem was subtle, "the more intangible consequence resulting from more detailed time-consuming control by the higher echelons outside of the Air Force." External controls removed "program control flexibilities" from "responsible operating levels in the field." Schriever's management reforms, encapsulated in the 375-series regulations, sought to decentralize decision-making processes to the project manager to reduce weapon system lead times.[81] He lamented the increasing delays and the volume of work generated by McNamara's requests for information, which subverted AFSC's procedures.[82]

Col. Otto Glasser of AFSC Headquarters summarized the new trends at a 1962 conference. He stated, "Dominant among all of the external influences in systems management is the trend toward centralized civilian control. . . ."[83] Due in large degree to the complexity and cost of new weapon systems, the "validation of requirements [for a weapon system] can no longer be left to the possible parochial opinions of a single service. . . . The requirements of today must be national and they must be validated in detail before a program can be seriously undertaken."[84] Upper management evaluated proposed systems primarily on the basis of cost-effectiveness. Once they had given their approval, they subjected programs to continual review.[85] As Glasser put it, the reforms "seem to be in the direction that you would go in handling your own household budget and I believe that it is safe to say that is just how the average taxpayer expects us to employ the tax dollars."[86] On the other hand, there were dangers. He compared current management to the old-style efficiency experts:

> The efficiency expert is not a creator, but rather an improver on other people's creations. For a limited period, the benefits of his labors can be enormous, but care must be taken to avoid stifling all new ideas. There can be no bold approaches when the program must be defined in explicit detail before initiation. If an approach must survive the review of a series of boards, panels, committees, advisors, etc., before it can be started, it is safe to say that the technology it employs will be the least common denominator among these many reviewers, and we risk the danger inherent in the story that "a camel is a horse designed by a committee."[87]

Schriever agreed: "If we are to be held to this overly conservative approach, I fear the timid will replace the bold and we will not be able to provide the advanced weapons the future of the nation demands."[88]

Schriever sensed the change in national priorities and saw the impact of McNamara's reforms. By 1962 studies by Harvard and RAND economists had shown that DOD weapons projects consistently had large overruns and schedule slips, with missile programs having the worst record. The RAND study showed that across six missile projects costs overran by more than a factor of four and schedules slipped over 50 percent. Other, nonmissile projects showed smaller slips, but these still averaged at least 70 percent cost overruns, with the average for all projects, missiles and others, exceeding 200 percent (triple the original cost estimates). With a record like that, the military was clearly vulnerable to criticism on cost issues.[89]

Replacing concurrency, managerial reform and cost control soon became Schriever's new watchwords. The immediate problem facing Schriever in early 1962 was to respond vigorously to the McNamara–Rubel initiatives, which he saw as cost control measures.[90] In a February 1962 memorandum to his commanding officer, General LeMay, Schriever stated that cost overruns arose from "any one or a combination of" factors, including deliberate underestimation, adhering to overly strict standards, overoptimism in estimating performance and schedules, vacillation or changes in program direction, and inadequate military or contractor management.[91]

One area that Schriever had to improve was cost estimation so he gave his comptroller's office the task. They began by educating his staff, instituting cost analysis training courses at the Air Force Institute of Technology (AFIT) in Dayton. By February 1962 the first class of 25 students graduated from the course. Schriever intended to train 125 more. AFSC also developed a new report, the "Program Planning Report." This report allowed for improved analysis of cost data with respect to technical and schedule progress and was "received enthusiastically" by the OSD and the Bureau of the Budget.[92]

To improve cost and schedule estimation, AFSC adopted and modified the Navy's new planning tool, the Program Evaluation and Review Technique (PERT). PERT used mathematical network techniques to connect scheduled events to estimate the overall schedule. Schriever "expended considerable effort to improve basic PERT. . . ." With "advanced PERT" he found that industry was "anxious to participate in order to improve its own management" in the use and improvement of PERT.[93] Figure 14 illustrates a generic PERT network.

Configuration control helped with cost estimation and with reliability, which was a major problem in missile, space, and electronic programs. Education was one of the primary means to attain higher reliability. AFSC enrolled fifteen officers in the master's program in reliability engineering at the Case Institute of Technology and in a course at AFIT. The Air Logistics School established a

Figure 14

Generalized PERT Network with Critical Path

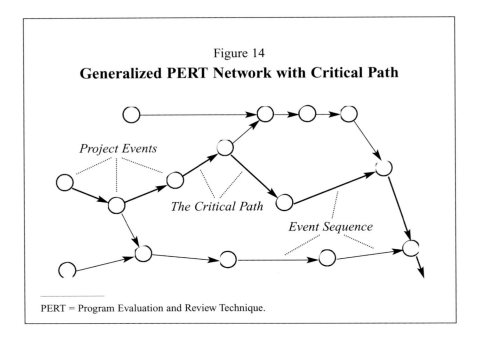

Project Events

The Critical Path

Event Sequence

PERT = Program Evaluation and Review Technique.

course in reliability, and more than 250 AFSC personnel completed a course in reliability from the Institute of Radio Engineers and the American Society for Quality Control. All new contracts required the use of the new reliability specification and AFSC reported progress in a semiannual Reliability Status Report. Enforcement of reliability standards emphasized the necessity of conservative designs in the McNamara regime.[94]

Schriever developed other ways to improve AFSC's management capabilities. He established a Management Improvement Board charged to "energize a vigorous program of management improvement and to provide overall guidance and direction to our management efforts." The board was "made up of General Officers having the greatest experience in systems management matters ranging from funding, systems engineering, procurement and production, through research and development." Schriever instructed it to examine "the entire area of systems management methods to include those of the Industrial complex as well as those of the Air Force." He also reinstated the Air Force Industry Advisory Group, a Board of Visitors to improve working relationships with industry, and a program of "systems management program surveys." Schriever claimed, "in a typical instance, one contractor has accepted and acted upon twenty-one recommendations made as a result of such a survey. These recommendations included changes to top-level organization, establishment of improved control procedures over scheduling, programming, budgeting, prime sub-contracting, and production planning." AFSC also collected "lessons learned" information from programs[95] and broadcast this information through publications and industry symposia,[96] and Schriever used the information to produce management goals for AFSC.[97]

AFSC communicated systems management concepts through several forms of education, including a new system program management course at AFIT[98] and a systems management newsletter within AFSC.[99] The AFIT course used case studies taught by experienced program managers, such as Brig. Gen. Samuel Phillips of the Minuteman program. Managers taught the concepts of program planning and budgeting, the McNamara reforms, organizational roles in system development, systems engineering, configuration management and testing, system acquisition regulations, program management techniques, contracting approaches, and financial methods.[100]

Despite his dislike for McNamara's centralizing changes, which moved decision making from the Air Force to the OSD,[101] Schriever emphasized management at AFSC, both for the better efficiency of AFSC itself and to improve the command's position as a management innovator. His initiatives resulted in the imposition of strong project management organizations in AFSC and in its industry contractors, in detailed "feedback and control" mechanisms used in day-to-day management, and in educational and communication mechanisms to transmit the methods to AFSC and industry. Schriever forcefully imposed his management methods throughout AFSC, and through its officers and publications he influenced many others.

McNamara, duly impressed with the procedures and reforms in Schriever's organization, used them in part as the basis for the DOD's new regulations for developing large-scale weapon systems. In 1965 the DOD enshrined phased planning and the system concept as the cornerstone of its R&D regulations. These processes, having already spread to NASA, moved throughout the aerospace and computing industries. Even when not used explicitly, the assumptions and ideas encompassed in these regulations became an accepted element of the culture of innovation in both venues.[102]

Conclusion

Despite the failures that plagued early intercontinental ballistic missile tests, Schriever's management of ballistic missiles development became a model widely known and often admired throughout the Air Force. As SAGE moved closer to its testing and operation, it faced difficult problems of organizational coordination to achieve the integration of sensors and weapons into the air defense network. Air Force leaders concerned with those problems and faced with the notorious independence of MIT's Lincoln Laboratory looked longingly to the installation at Inglewood for ideas on how to solve their problems. Civilians at Lincoln, however, did not share the Air Force's views because, despite the problems of informal coordination, they believed that the independence of civilian organizations was a necessary ingredient to success, as it had been for SAGE.

Thus, in the organizational crisis caused by MIT's desire to extricate Lincoln from large-scale development, two distinct threads of thought emerged. One line of thought, shared by Lincoln Division 6 leaders such as Robert Everett and experienced air defense leaders like Maj. Gen. Kenneth Bergquist, was that the U.S. air defense systems required extensive coordination across U.S. and Canadian armed services and with operating commands such as ADC. The ideal organization would be at highest level, either with NORAD or the DOD, to enforce cross-service decisions. In that view, explicit and detailed interactions with the commands that would use the systems were critical during the entire development process, because automation of command and control required an in-depth understanding of current operations, tactics, and strategy. As technologies developed, so, too, would operations, requiring further changes to the technologies. In essence, this was an "evolutionary" model of incremental development, with organizational solutions taking precedence over technical ones.

The opposing view was that the Inglewood model of technological development should be adopted throughout the Air Force, including electronic systems. In that view, extensive up-front planning at the Air Staff level would set top-level system requirements, and an "empowered" project office would be given the

MITRE president C. W. Halligan (center) in conference with NORAD Commander Gen. Laurence S. Kuter (left) and ESD Commander Maj. Gen. Kenneth P. Bergquist. Bergquist unsuccessfully lobbied to place ESD and MITRE in the Department of Defense or NORAD.

authority to develop the system. By this process, the end-user commands had some influence in setting the initial requirements, but their influence during development was limited. Authority in the ICBM case came from the top level, but in theory this did not need to be the case if the Air Force delegated substantial authority to the project officers to manage their systems. That was a "revolutionary" model of technical development in which the technologies took precedence and organizational adjustments to operate them came later.

Ironically, the Inglewood model, which did not necessarily require authority at the very top for implementation, had this authority for ICBMs. By contrast, the Lexington model for air defense needed this authority to operate properly but never got it. Those contradicting realities, influenced primarily by political and strategic decisions at the highest levels, confused the debate about the best way to improve R&D across the Air Force.

In the end, of course, Schriever's Inglewood model triumphed for a number of reasons. One was undoubtedly the strong, politically savvy leadership of Schriever himself. What officer failed to admire what Schriever had done in creating a devastating new weapon while carving out a bureaucratic empire answerable to no one except the Congress and the president? By contrast, no powerful military leader took air defense so seriously because of the "offensive bias" of the Air Force, which valued better ways to deliver destructive force to the Soviet Union over the possibilities of improved air defense. That bias, traceable to the hard lessons of World War II, loomed large for most of the Air Force's leaders. Thus the lessons of Inglewood were far better known and understood than those of Lexington. Nor were Air Force leaders willing to stick their necks out for air defense like Schriever had done for ICBMs. It just didn't seem to be worth it. Finally, there was the highly visible nature of ICBMs in contrast to command and control. Success (and failure) were readily apparent and easily defined for ICBMs: the proper size explosion within a certain distance of the target. For command and control, success and failure were more intangible and subjective. By what criteria could SAGE be declared a success? Some argued that SAGE was "a lemon," and others that it was a technological and operational marvel. Either point could be (and both were) argued, but neither was proven.

When Schriever became head of ARDC, reorganization based on the Inglewood model was a foregone conclusion. The sweeping changes affected processes through the 375 Procedures and the organization structure, with the formation of ASD, BSD, and ESD. Those changes brought ballistic missiles back into the Air Force's normal structure, gradually replacing DOD-level approvals with Air Force processes. In this reorganization the Air Force eliminated the independence of both Ramo-Wooldridge, with the spin-off of Aerospace Corporation, and MITRE. Both became nonprofit contractors that handled those technical details that the Air Force assigned to them. The Air Force effectively quashed independent planning and initiatives within both organizations. Within ESD greater involvement of the end-user commands and evolutionary planning soon found

Gen. Bernard Schriever, commander of the Air Force Systems Command, met with key personnel of the Ballistic Systems Division at Norton Air Force Base in 1964. Pictured at the table (left to right) are Al Donovan, Maj. Gen. Ben Funk, Maj. Gen. Austin Davis, General Schriever, Brig. Gen. John L. McCoy, and Ivan Getting.

their way back into the organization because optimum technical development truly required them.

The Air Force's organizational reforms caught the attention of arch-manager Robert McNamara in the early 1960s. Whereas Schriever's centralization aimed to speed development and give authority to project managers, McNamara used centralization to control costs and reduce duplication. Much to Schriever's dismay, McNamara found the Inglewood model congenial for it developed the communication and control methods necessary to control technical development. McNamara needed only minor modifications such as phased planning to transform the Inglewood model into a centralized planning and budgeting system across the services. Systems management thus spread through all of the services, tying proven R&D methods to civilian managerial control. With systems management the DOD now could plan and control technical development for its own ends.

Notes

1. John F. Jacobs, *The SAGE Air Defense System: A Personal History* (Bedford, Mass: MITRE Corporation, 1986), pp 164–165.

2. Author unknown, "Outline of SDC History (I–IV)," June 15, 1964, "SDC Hist (Including STP, PACOM, Command Research) 1960–1964" folder, SDC

series, box 1, Burroughs Collection, CBI 90; F. N. Marzocco, "The Story of SDD," SD-1094, Oct 1, 1956, p 1, "RAND Corp. SDD Hist 1956–7" folder, box 1, SDCP, Burroughs Collection, CBI 90, pp 7, 8; Claude Baum, *The System Builders: The Story of SDC* (Santa Monica: SDC, 1981), pp 26–28.

3. Thomas P. Hughes, *Rescuing Prometheus* (New York: Pantheon, 1998), p 62.

4. Minutes, Lincoln Steering Cmte mtg, Oct 31, 1955, MITRE, AC-22, p 2.

5. Jacobs, *SAGE Air Defense System,* p 130; ltr, Lt Gen D. L. Putt, DCS/D, to Lt Gen Thomas S. Power, Commander, ARDC, Feb 20, 1956.

6. Richard F. McMullen, *The Birth of SAGE 1951–1958,* ADC Historical Study no 33, Lincoln, accession #358587, pp 43–46

7. Memo 6M-4048, Jay W. Forrester to distribution list, subj: General Procedure for Integrating New Weapons with SAGE," Jan 30, 1956, p 1.

8. Ltr, Brig Gen Ivan L. Farman, Asst for Electronic Supporting Systems, to RDZS, RDZE, subj: Integration of SAGE/Air Defense Weapon Systems, MITRE, box 1782.

9. Memo, Col Albert R. Shiely, Jr., Chief, Air Defense Systems Operating Div, to Col G. T. Gould, Jr., and Lt Col G. J. Watts, Dec 15, 1955, MITRE, box 1773.

10. *Ibid.*

11. Ltr, Maj Gen William F. McKee, Vice Commander, AMC, to DCS/D, Mar 27, 1956, MITRE, box 1782.

12. Ltr, Col Albert R. Shiely, Jr., Chief, Air Defense Systems Operating Div, to Dr. George Valley, Assoc Dir, Lincoln, Apr 20, 1956, MITRE, box 1782; memo 2C-0568, R. E. Rader to distribution list, subj: Minutes of ADES–Lincoln Engineering Mtg, Jun 27, 1956, MITRE, box 1783.

13. Howard R. Murphy, *The Early History of the MITRE Corporation: Its Background, Inception, and First Five Years* (Bedford, Mass: MITRE Corporation, 1972), pp 66–68.

14. *Ibid.*

15. Jacobs, *SAGE Air Defense System,* pp 139–141; Hughes, *Rescuing Prometheus,* pp 62–63; Murphy, *Early History of the MITRE Corporation,* pp 63–64.

16. Jacobs, *SAGE Air Defense System,* pp 137–138; Murphy, *Early History of the MITRE Corporation,* pp 64–65.

17. Ltr, R. R. Everett to C. F. J. Overhage and V. H. Radford, subj: Air Defense Systems Engineering, Jun 3, 1957, MITRE, box 1782.

18. Murphy, *Early History of the MITRE Corporation,* pp 65–66.

19. *MITRE, The First Twenty Years, A History of the MITRE Corporation (1958–1978)* (Bedford, Mass: MITRE Corporation, 1979), pp 12–14; Murphy, *Early History of the MITRE Corporation,* pp 74–78.

20. *MITRE: The First Twenty Years,* pp 13–18; 2d draft memo, R. R. Everett and J. F. Jacobs to Lincoln Steering Cmte, subj: A Proposal for Providing Systems Engineering Services to ADSMO, Aug 5, 1957, MITRE, Jacobs Collection, AC-61.

21. AFR 70-9, Procurement and Contracting, Nov 12, 1953, MDP, AFHRA, 168.7265-237; US Cong, House Cmte on Government Ops, 87th Cong, 1st Sess, 3d Rpt, AF Ballistic Missile Management (Formation of Aerospace Corporation), rpt no 324 (Washington, D.C.: GPO, 1961), pp 4–7.

22. Ethel M. DeHaven, *AMC Participation in the AF Missiles Prog through Dec 1957* (Inglewood: Ofc of Information Services, AMC Ballistic Missile Ofc), AFHRA, roll 2312, pp 184–186.

23. Davis Dyer, *TRW: Pioneering Technology and Innovation Since 1900* (Boston: Harvard Business School Press, 1998), pp 198–206.

24. *Ibid.*, pp 218–221.

25. *Ibid.*, pp 228–232; US Cong, House Cmte on Government Ops, Formation of Aerospace Corporation, pp 12–18.

26. The Directorate of Systems Management was a HQ ARDC organization founded just prior to this time, and it transferred major systems planning and

programming (funding allocation) from AF field centers to HQ; see Harry C. Jordan, *History of BSD, AFSC, The Div Pronaos,* Historical Div, Ofc of Information, BSD, AFSC, Norton AFB, Calif, Apr 3, 1963, MITRE, box 1753, pp 18–19.

27. Minutes, Conference on Continental Air Defense, HQ ARDC, Nov 19, 1956, in Kent C. Redmond and Harry C. Jordan, *Air Defense Management, 1950–1960, The ADSID,* vol II, part A, 1961, ARDC Historical series: 61-31-I, HRL.

28. Minutes, Conference on Continental Air Defense, HQ AMC, Dec 6, 1956, in Redmond and Jordan, *Air Defense Management,* vol II; "Proposal for the Establishment of the Air Defense Management Ofc," Dec 14, 1956, in Redmond and Jordan, *Air Defense Management,* vol II.

29. Jacobs, *SAGE Air Defense System* (Bedford, Mass: MITRE Corporation, 1986), p 135; Brig Gen Donald R. Hutchinson, Air Defense Management Systems, presentation to the USAF Air Council, Apr 15, 1957, in Redmond and Jordan, *Air Defense Management,* vol II; Brig Gen D. R. Hutchinson, Improvement of Air Defense Systems Management, rpt to the Secretary of the AF, Apr 17, 1957, in Redmond and Jordan, *Air Defense Management,* vol II.

30. Ltr, ARDC Commander to Nathan Twining, CofS, Apr 25, 1957, in Redmond and Jordan, *Air Defense Management,* vol II.

31. *Ibid.*

32. Briefing, ADSMO to Gen Rawlings, Lt Gen Anderson, Lt Gen Atkinson, Aug 1957, in Redmond and Jordan, *Air Defense Management,* vol II.

33. Jacobs, *SAGE Air Defense System,* pp 135–136.

34. Howard R. Murphy, *History of the AFCCDD, 16 Nov 1959–31 Mar 1961,* vol I, prologue to the Hanscom Complex, Historical Div, Ofc of Information, ESD, Hanscom Field, Bedford, Mass, AFSC, Jul 1964, AFSC Historical Publication 64-32-I, pp II-44–II-46; memo, Maj Gen Howell M. Estes, Jr., Asst CofS for Air Defense Systems, to the Comptroller of the AF, subj: Management of Air Defense Systems, Feb 24, 1958, in Redmond and Jordan, *Air Defense Management,* vol II; Kent C. Redmond and Harry C. Jordan, *Air Defense Management, 1950–1960: The ADSID,* vol I, Historical Branch, Ofc of Information, AFCCDD, Laurence G. Hanscom Field, Bedford, Mass, Feb 1961, pp 47–50.

35. Jacobs, *SAGE Air Defense System,* p 141; Redmond and Jordan, *Air Defense Management,* vol I, pp 48–49.

36. Murphy, *History of the AFCCDD,* pp III-68–III-70.

37. Redmond and Jordan, *Air Defense Management,* vol I, pp 68–72.

38. Dennis J. Stanley and John J. Weaver, *An AF Command for R&D, 1949–1976: The History of ARDC/AFSC* (Washington, D.C.: Ofc of Hist, HQ AFSC, 1976), p 31.

39. Michael H. Gorn, *Vulcan's Forge: The Making of an AF Command for Weapons Acquisition (1950–1985)* (Washington, D.C.: Ofc of Hist, HQ AFSC, 1989), pp 53–56.

40. *Ibid.*, p 57; Murphy, *History of the AFCCDD,* pp III-72–III-73.

41. Gorn, *Vulcan's Forge,* pp 59–61; intvw, Gen Bernard A. Schriever with author, Apr 13, 1999, USAF HSO.

42. Gorn, *Vulcan's Forge,* pp 63–64; Murphy, *History of the AFCCDD,* pp III-74–III-75.

43. Murphy, *History of the AFCCDD,* pp III-75–III-83.

44. *Ibid.*; *MITRE: The First Twenty Years,* pp 33–36.

45. Murphy, *History of the AFCCDD,* pp III-85–III-92.

46. Intvw, Everett, Oct 13, 1998.

47. Murphy, *History of the AFCCDD,* pp IV-5–IV-22, VI-118–VI-122.

48. *Ibid.*, pp IV-11–IV-13.

49. MITRE Corporation, "Winter Study Rpt: The Challenge of Command and Control," Mar 31, 1961, MITRE.

50. Murphy, *History of the AFCCDD,* pp VI-126–VI-129.

51. *Ibid.*, pp VI-127–VI-134.

52. Murphy, *Early History of the MITRE Corporation,* pp 219, 250–253; memo, O. T. Halley, Jr., to distribution list, subj: ESD/MITRE Relationships, Sep 11, 1962, MITRE, box 1783.

53. Gorn, *Vulcan's Forge,* pp 63–69; Stanley and Weaver, *AF Command for R&D,* pp 34–37. The most detailed description of the entire Anderson Group effort is in Jordan, *History of BSD.*

54. See intvw, Schriever, Apr 13, 1999; Gorn, *Vulcan's Forge,* p 69; Stanley and Weaver, *AF Command for R&D,* p 37; Jordan, *History of BSD,* p 97.

55. Dept of the AF, *AFR No 375-1, Systems Management, Management of Systems Progs,* Feb 12, 1962, LOC/SPP, box 37, p 1.

56. Dept of the AF, *AFR No 375-2, Systems Management, System Prog Ofc,* Feb 12, 1962, LOC/SPP, box 37.

57. Dept of the AF, *AFR No 375-3, Systems Management, System Prog Dir,* Feb 12, 1962, LOC/SPP, box 37.

58. Dept of the AF, *AFR No 375-4, Systems Management, System Prog Documentation,* Feb 12, 1962, LOC/SPP, box 37.

59. Dept of the AF, *AFR No 80-5, Research and Development, Reliability Prog for Systems, Subsystems, and Equipments,* Jun 4, 1962, LOC/SPP, box 37.

60. See Gorn, *Vulcan's Forge,* pp 71–72; Stanley and Weaver, *AF Command for R&D,* pp 39–41; David N. Spires, *Beyond Horizons: A Half Century of AF Space Leadership* (Washington, D.C.: GPO, 1997), pp 86–92.

61. Gorn, *Vulcan's Forge,* p 72; Stanley and Weaver, *AF Command for R&D,* pp 41, 44.

62. Address, Maj Gen O. J. Ritland, AFSC Management Conference, Monterey, Calif, May 2, 1962, AFHRA, microfilm 26254, pp 26–27.

63. USAF, AFSC, *Rainbow Reporting: Systems Data Presentation and Reporting Procedures, Prog Management Instruction 1–5,* Jan 15, 1963, LOC/SPP, box 18.

64. Memo, Configuration Control Group to Col K. N. Retzer, subj: Data Processing Equipment for Phase C, Oct 25, 1961, MITRE, AC-60; memo M/C-402, J. B. Frazer to C. C. Grandy, subj: SPADATS Configuration for Phase B (496), Nov 17, 1961, MITRE, AC-60; working paper W-5267, C. G. Teschner, M. G. Murley, J. I. Richter, subj: 425L Configuration Management Instructions, Aug 21, 1962, MITRE, AC-60.

65. See Jacob Neufeld, ed, *Reflections on R&D in the USAF: An Interview with Gen Bernard A. Schriever and Gens Samuel C. Phillips, Robert T. Marsh, and James H. Doolittle, and Dr. Ivan A. Getting,* conducted by Dr. Richard Kohn (Washington, D.C.: Ctr for AF Hist, 1993), p 65.

66. C. W. Borklund, *The DOD* (New York: Frederick A. Praeger, 1968), pp 39–46.

67. *Ibid.,* pp 50–56.

68. *Ibid.,* pp 57–67; John J. McLaughlin, "The AF's Management Revolution," *AF Magazine* (Sep 1962), pp 98–106.

69. See Deborah Shapley, *Promise and Power, The Life and Times of Robert McNamara* (Boston: Little, Brown and Company, 1993).

70. Borklund, *DOD,* pp 70–71; Charles J. Hitch, *Decision-Making for Defense* (Berkeley: University of Calif Press, 1967), pp 31–32; memo, F. T. Moore to Research Council, subj: "The 92 Labors of Secretary McNamara," March 17, 1961, WM-663, CBI 90, Burroughs Collection, SDC series, box 1, "RAND Corporation: Correspondence and Clippings," folder.

71. Shapley, *Promise and Power,* pp 99–101.

72. James M. Roherty, *Decisions of Robert S. McNamara* (Coral Gables: University of Miami Press, 1970), pp 72–73.

73. Internal summ, "DOD System Acquisition Policies," used for a mtg in the AF, ca 1962, LOC/SPP, box 19.

74. Memo, John H. Rubel, Dep Dir, DDRE, to Secretary of Defense, subj: Management of Research and Engineering, Nov 14, 1961, LOC/SPP, box 19.

75. Memo, John H. Rubel, Dep Dir, DDRE, to Asst Secretary of the AF (R&D), subj: Standardized *AGENA,* Oct 4, 1961, LOC/SPP, box 19.

76. Memo, John H. Rubel, Dep Dir, DDRE, to Asst Secretary of the AF (R&D), subj: Titan III Launch Vehicle Family, Oct 13, 1961, LOC/SPP, box 19.

77. Memo, John H. Rubel, Dep Dir, DDRE, to Asst Secretary of the AF (Material) and the Asst Secretary of the AF (R&D), subj: Planning for Titan III Phase I Efforts, Dec 6, 1961, LOC/SPP, box 19.

78. Memo, John H. Rubel, for Harold Brown, Dir, DDRE, the Secretary of the Navy and the Secretary of the AF, subj: Mobile Mid-Range Ballistic Missile Prog Plan, Feb 19, 1962, LOC/SPP, box 19.

79. Memo, John H. Rubel, Dep Dir, DDRE, to the Asst Secretaries of the Army, Navy, and AF (R&D), subj: Management of Research and Engineering, Oct 9, 1961, LOC/SPP, box 19.

80. Memo, Col Jewell C. Maxwell, USAF, Asst for Management Policy, AFSC, to HQ USAF (AFSDC), subj: Systems Management and Prog Initiation, May 9, 1962, LOC/SPP, box 19; see also Michael E. Brown, *Flying Blind: The Politics of the US Strategic Bomber Program* (Ithaca: Cornell University Press, 1992) on concurrency vs sequential procurement. This episode contradicts Brown's view that the McNamara years were the height of concurrency because here we see McNamara pressing for more sequential procurement to plan for costs more efficiently.

81. Ltr, Gen B. A. Schriever, USAF/AFSC, to Gen C. E. LeMay, HQ USAF, Apr 30, 1962, LOC/SPP, box 19.

82. *Ibid.,* p 1.

83. Col Otto J. Glasser, HQ USAF, AFSC, "Current AFSC Management Environment," ca 1962, LOC/SPP, box 19, p 5-1-1.

84. *Ibid.,* p 5-1-3.

85. *DOD Annual Rpt for FY 1961,* quoted from Glasser, "Current AFSC Management Environment," p 5-1-5.

86. Glasser, "Current AFSC Management Environment," pp 5-1-7, 5-1-8.

87. *Ibid.*

88. Ltr, Schriever to LeMay, Apr 30, 1962, p 4.

89. See Merton J. Peck and Frederic M. Scherer, *The Weapons Acquisition Process: An Economic Analysis* (Boston: Div of Research, Graduate School of Business Administration, Harvard University, 1962), pp 19–24; Ants Kutzer, "Implementation of a Satellite Project," paper presented on the occasion of leaving the European Space Tech Ctr on Sep 15, 1967, HAEUI Fond 51048, table I.

90. Memo, Gen B. A. Schriever, Commander AFSC, to Gen C. E. LeMay, HQ USAF, subj: Systems Acquisition Management Improvement, Feb 5, 1962, LOC/SPP, box 19

91. *Ibid.,* p 1.

92. Memo, Gen B. A. Schriever, HQ AFSC, to Gen C. E. LeMay, HQ USAF, subj: Systems Acquisition Management Improvement, Feb 5, 1962, LOC/SPP, box 19.

93. *Ibid.,* p 2; Special Projects Ofc, "Prog Evaluation Research Task, Summ Rpt, Phase 1," Bureau of Naval Weapons, Dept of the Navy, Washington DC, Jul 1958. It was later renamed Prog Evaluation and Review Technique.

94. Memo, Schriever to LeMay, Feb 5, 1962, p 3.

95. USAF, AFSC, *A Summary of Lessons Learned from AF Management Surveys,* AFSCP 375-2, Jun 1, 1963, LOC/SPP, box 36.

96. Memo, Maj Gen Robert J. Friedman, DCS/Comptroller, to distribution list, subj: Guidance and Plan for Follow-Through on the Work of the Monterey Management Conference, Jun 12, 1962, LOC/SPP, box 19.

97. USAF, AFSC, *AFSC Management Objectives, FY 1964,* Aug 1963, LOC/SPP, box 18.

98. See viewgraphs for AFIT System Prog Management Course, probably presented by Samuel C. Phillips, ca 1963, LOC/SPP, box 37.

99. USAF, AFSC, *Systems Management Newsletter,* no 6, May 1963, LOC/SPP, box 18.

100. Memo, Maj Gen Samuel C. Phillips, Dep Dir, Apollo Prog, to MSFC, KSC, MSC, subj: System Prog Management Course at AFIT, Oct 10, 1964, LOC/SPP, box 37.

101. Schriever later noted that when McNamara took office, the days when he could "get things done" were over. He thought the bureaucracy removed flexibility and speed from the military. See intvw, Gen Bernard A. Schriever with Dr. Edgar F. Puryear, Jr., USAF Oral Hist Prog, Jun 15 and 29, 1977, AFHRA, K239.0512-1492, p 7; see also Neufeld, ed, *Reflections*, pp 67–78.

102. DOD Directive 3200.9, "Initiation of Engineering and Operational System Development," Jul 1, 1965.

Chapter 6

Securing the Technological Future

> Leadership in the development of these new weapons of the future can be assured only by uniting experts in aerodynamics, structural design, electronics, servomechanisms, gyros, control devices, propulsion, and warhead under one leadership, and providing them with facilities for laboratory and model shop production in their specialties and with facilities for field tests. Such a center must be adequately supported by the highest ranking military and civilian leadership and must be adequately financed. . . .

> Theodore von Kármán, 1946[1]

Since its inception the United States Air Force has depended on advanced technologies to maintain an edge over its actual and potential enemies. Continuous innovation became a way of life and Air Force leaders learned quickly to foster productive relationships among their service and the scientists, engineers, and industry leaders who built the aircraft, missiles, computers, radar systems, and other technologies on which the Air Force depended. When we consider the criticality of advanced technology for air and space warfare, it is not terribly surprising that the Air Force created and standardized organizational means to foster technical creativity and harness it into offensive and defensive weaponry. What is more surprising, perhaps, is that criticism of Air Force methods for technology development frequently has been so severe. By the only yardstick that really counts for the military—that is, in combat—the Air Force maintained air superiority in every war fought by the United States from World War II to the end of the twentieth century.

Much of the methods' criticism can be attributed to measurement against a different standard: economics. Just as military effectiveness is best assessed by actual combat, economic effectiveness is best assessed by the competitive performance of the affected industries in international competition. It so happens that two of the American industries most affected by Air Force spending and methods—aerospace and computing—have been the most successful of all United States industries in international competition. Is this merely a coincidence? Probably not.

Of the three military services, the Air Force relied (and continues to rely) most heavily on academia and industry. Instead of bottling up technologies and the methods for developing them in a government arsenal, the Air Force created methods that explicitly crossed organizational boundaries from the government to industry and academia. By 1965 the Department of Defense as a whole standardized these methods across the entire department, and NASA, too, had adopted them. Although it is beyond the scope of this book, it is a fact that by the late 1960s these methods dominated aerospace and, to a lesser degree, computing. They formed the organizational culture and influenced expectations of the managers, scientists, and engineers in those industries and in the academic disciplines that supported them.

Recruiting the Scientists and Engineers

To work effectively with scientists and engineers in developing new technologies, the Air Force devised organizational structures and processes to facilitate interactions among Air Force officers on one side and scientists and engineers on the other. Immediately following World War II, most scientists resided at universities, including such influential veterans of World War II efforts as George Valley at MIT and John von Neumann at Princeton. Engineers could be found both in universities, like Jay Forrester at MIT and Theodore von Kármán at the California Institute of Technology, and in industry, such as Hughes Aircraft's Simon Ramo and Dean Wooldridge.

The Air Force had existing mechanisms for working with engineers and managers in the aircraft industry, but not methods that worked with new technologies such as nuclear weapons, missiles, radar, operations research, computers, and rockets. For those technologies the Air Force had to seek expertise wherever it could be found because the aircraft industry's capabilities in those areas were limited. The Douglas Aircraft Company spun off its operations research capability into RAND Corporation, the first nonprofit corporation for military purposes. RAND quickly expanded its capabilities to include scientists from many disciplines, and its unique relationship with the Air Force gave it privileged access.

Army Air Forces Chief Henry "Hap" Arnold established the scientific liaison position, filled first by Col. Bernard Schriever, to enhance communications among scientists and the Air Force's senior leaders. The Scientific Advisory Board came into being for exactly the same purpose. The Research and Development Board had a coordination and communication function among all of the services and the technical experts. Scientists and engineers, of course, favored increased funding for and attention to scientific and technological research. They were natural allies of military officers such as Arnold and Brig. Gen. Donald Putt, who believed that the Air Force should create a separate organization for R&D. The SAB's 1949 Ridenour Report led directly to the 1950 creation of such an organization, the Air Research and Development Command.

Much attention focused on the development of research capabilities in the late 1940s but by the early 1950s the Air Force's attention increasingly converged on technology development. Military and civilian research had led to the creation of radar, rockets, computers, and nuclear weapons. By 1950 each of those new technologies had developed sufficiently to be used as part of larger systems, such as ICBMs and continental air defense networks.

Lt. Gen. Donald L. Putt was a proponent of technical innovation.

Development increasingly took center stage. Jay Forrester's research computer caught the attention of MIT physicist George Valley in 1950 as Valley searched for an air defense computer. Similarly, when physicists such as Edward Teller and John von Neumann demonstrated nuclear fusion weapons (hydrogen bombs) in 1952 and 1953, Convair's experimental rocket, the Atlas, suddenly became a plausible candidate for serious development. Research scientists and engineers unquestionably took the lead in developing the early prototypes. The question was who would lead the full-scale development of technological systems that used more advanced versions of these technologies?

In air defense, the Air Force's relative lack of interest and the strong interest of Radiation Laboratory veterans such as George Valley created a situation in which MIT took the lead. Firmly believing in the effectiveness of offense over defense, senior Air Force leaders supported development of early warning and air defense as long as it did not detract from nuclear weapon and bomber development. Preferring to commit their time and effort to bombers and later missiles, the Air Force's senior officers were satisfied to let MIT's experienced scientists and engineers develop the air defense system. Because the government had the official authority but lacked the expertise or interest, all parties concerned had to tread carefully around the reality of MIT's leadership.

MIT committed to SAGE development partly out of a sense of duty, but also out of a realization that SAGE could be a landmark in technology development.

That intuition was correct because SAGE and Lincoln Laboratory became critical elements contributing to MIT's technical leadership. However, as SAGE approached operations by 1956 and 1957, MIT's leaders began to feel the negative ramifications of large-scale development.

Lincoln's model of an academic institution managing large-scale development for the military was not stable over the long run and MIT's leaders maneuvered to return Lincoln to free-ranging research. The military needed an organization under their supervision that could help them develop large-scale electronic systems. Lincoln had the expertise but it demanded too much independence. Large-scale development with its massive funding required much closer military supervision. Furthermore, potential conflicts of interest loomed with industry if Lincoln became a competitor for system development. The solution to those problems was the nonprofit corporation MITRE. Not connected to MIT, MITRE did not need to worry about competing with industry. The Air Force preferred MITRE because it could control the corporation much more easily than it could the stubbornly independent and politically powerful Cambridge professors.

Ballistic missiles presented a somewhat different situation. The Air Force's need was the same—knowledgeable scientists and engineers to guide the development of ballistic missiles—but the Air Force, and particularly General Schriever, were going to call the shots. There was no academic institution available to run the ICBM effort for the Air Force; the only qualified facility, the California Institute of Technology's Jet Propulsion Laboratory, worked for Army Ordnance. The next-best option was a small company in the right place at the right time, Ramo-Wooldridge. Simon Ramo and Dean Wooldridge were highly educated electronics and missile experts from Hughes Aircraft who espoused the systems approach that they had successfully applied on the innovative Falcon missile at Hughes.

Both Schriever and the Teapot Committee headed by von Neumann were convinced that Convair did not have the appropriate expertise. Furthermore, both believed that physical scientists needed to direct the ICBM effort, just as they had done for the World War II Manhattan Project. R-W was a small company with respected leaders searching for business. The principals were not particularly eager to take the job but at Schriever's insistence and warnings that they would not get Air Force contracts if they refused, they accepted the task. After hiring Jet Propulsion Laboratory director Louis Dunn to run the Atlas effort, R-W then hired the highly trained scientists and engineers that Schriever wanted.

R-W successfully assisted Schriever and the ICBM program but it ran into political trouble because of its intimate ties with the Air Force. Correctly fearing a powerful new competitor, other aircraft companies led by Convair waged an unrelenting political campaign against this "unnatural" relationship between the Air Force and a profit-making corporation. Despite their joint technical success, in the end neither the Air Force nor R-W could justify this violation of contractor propriety. As with Lincoln Laboratory, the solution for R-W and the Air Force was to create another nonprofit corporation to provide systems engineering for the Air Force—the Aerospace Corporation headed initially by Radiation Laboratory veteran and Schriever ally Ivan Getting. Ultimately, potential competition with industry for large development contracts militated against both MIT and R-W's too-close relationship to the government.

Solving the Puzzle of Complexity

For both ballistic missiles and command and control systems, technical complexity was the underlying problem that required scientific and engineering expertise. Despite the seeming dissimilarity, both kinds of systems shared common characteristics of technical breadth and depth. Both required the integration of various technical disciplines. For ballistic missiles, these included rocket propulsion, structures, electronics, guidance and control, and aerodynamics. For C^2 systems, knowledge of radar, computers, communications gear, and software dominated. Software, however, could not be built without an in-depth knowledge of air defense tactics. Depth of knowledge was necessary in both systems and so required training in mechanical and electrical engineering, propulsion, materials, aerodynamics, and military tactics.

Reliability was a second important technical issue for C^2 systems and ballistic missiles. The SAGE system had to be orders of magnitude more reliable than other computer systems of the 1950s. Whereas other computers were considered reliable if they worked for several hours at a time without failure, SAGE had to operate with just a few hours of downtime over an entire year. Ballistic missiles that delivered nuclear warheads also had to be extremely reliable for failure resulted in catastrophic explosions. Because each missile flew only once and never returned, it had to work the first time. The same was also true for spacecraft. "Getting it right" was critical for all of these technologies, such that miscommunication could not be tolerated. Formal documentation of methods, processes, and designs was important to ensure that everyone clearly understood everyone else, and to ensure that the thousands of components that went into these systems connected properly and were each highly reliable.

These technical problems alone pointed to the creation of the technical generalist and coordinator function that was soon known as "systems engineering." Systems engineering, however, received another spur from the complex organizational relationships involved with these systems. Communications among scientists, engineers, military officers, and managers were not easy because each of them spoke a somewhat different vernacular and had varied interests. In addition, they resided in different kinds of organizations, including for-profit and nonprofit corporations, military and civilian government agencies, and academic departments and semi-independent academic laboratories. Communicating across these numerous boundaries required an effort and attention to detail unnecessary in a more homogeneous organizational or cultural milieu. Legal and contractual requirements intervened as well, leading to greater formality than was necessary in a single organization.

Together, these technical and social considerations led to the creation of systems engineering as a new function within complex projects. Systems engineers were the "jacks of all trades"—people who acted as bridges between the different cultures and organizations. Then, as now, systems engineers typically were not trained as such in universities, but rather gravitated to this position through experience by dint of having good technical and communication skills. They became the technical partners of project managers, translating managerial edicts into technical terms and technical issues into terminology understandable to managers.

Communication skill by itself was not sufficient for managers to gain control over unruly scientists and engineers. Through the 1950s management literature was replete with complaints about the difficulties of working with technical experts, and with concerns that no methods existed for predicting or controlling the process or its products. With the development of systems engineering, technical communication was becoming more visible. The key to managerial control was to connect this technical information with things that managers could control, such as budgets.

Configuration management was the managerial solution for controlling technical development. By the late 1950s, for both C^2 systems and ballistic missiles, engineers had developed and were using methods of change control to coordinate engineering modifications. Change control generally required a committee or project manager to review and coordinate all changes to a given design. This in turn required those who requested changes to document the technical aspects of those changes. It was not a big step for the project manager then to require a cost and schedule estimate along with the technical description before approving a change. Each change, when approved, triggered changes to appropriate specification and design documents along with the hardware and software. This additional information altered change control into configuration management.

The importance of configuration management generally has not been recognized. It is critical for two major reasons. First, the configuration control board serves as the primary link between managerial hierarchy and engineering work groups because both are represented on the board. Second, configuration control is the primary mechanism for management to predict and control costs in development projects, and hence is also the primary means to control scientists and engineers. After the implementation of configuration management, scientists and engineers no longer could change the design and run up costs based on their local judgments. Management now approved or vetoed all changes and could modify the predicted funding requirements based on the documented decisions. The importance of cost prediction can hardly be overestimated for it is the basis of resource allocation and funding requests. Configuration management has been at the heart of aerospace and software engineering from the late 1950s to the present.

Reconciling Political Interests

At Inglewood and at Hanscom Field, where the electronics efforts ultimately centered, scientists and engineers created technical and organizational methods to coordinate and control complex technology development. These methods, however, had to be reconciled with the values and interests of the times and with the political and social issues at stake. One clear-cut conflict already mentioned was the need to convert the R-W and Lincoln systems engineering efforts into the nonprofit corporations Aerospace and MITRE.

The overriding political issue of the late 1940s through the 1950s was the Cold War between the United States and the Soviet Union. Stalin's version of Communism was a direct threat to capitalism and democracy and made national security the overriding priority of the U.S. government throughout the period. In the emergencies following the Soviet tests of the fission bomb in 1949 and the fusion bomb in 1953, military and political leaders looked to physicists and mathematicians to bring technological solutions to the Soviet military threat. In that atmosphere, scientists had a free hand and Air Force support to develop the necessary countermeasures.

In Cambridge, MIT's computing and radar experimentation that led to SAGE created an obvious reliance of Air Force leaders on MIT Radiation Laboratory veterans like George Valley. Scientists had a somewhat different role in the development of ICBMs. Schriever depended on John von Neumann and others to help the Air Force organize and direct ICBM development. The recommendations of the von Neumann Committee confirmed Schriever's opinion that Convair was not qualified to develop Atlas, and that R-W should coordinate the task. R-W could hire the necessary scientific and technical experts, such as Jet Propulsion Laboratory's Louis Dunn.

The Air Force's bias toward offensive weaponry over defense systems helped ensure the dichotomy between operations at Hanscom and Inglewood. MIT would have a free hand for the relatively unimportant SAGE system, but Schriever firmly would direct ICBMs. This mirrored the situation with research and development as a whole: scientists were free to direct their own—relatively unimportant—research, but the military generally took firm control over important and expensive development.

By the early 1960s the Cold War's initial heat began to fade and the aura of centralized management was glowing more brightly. The military's own experiences enhanced this shift as the revolutionary new systems went through their teething pains in full-scale tests in the late 1950s. Early test failures—a common feature of new product development—gave the appearance of incompetence. At the same time, systems engineering and the new technological marvels also led to the belief that scientists, engineers, and managers could solve virtually any problem. McNamara's managerial biases combined with the Eisenhower reforms that gave the Office of the Secretary of Defense significant power led to the priority of cost and schedule concerns over the technical performance issues that dominated the 1940s and 1950s. All of that coincided with the dominance of scientific values in the 1940s and 1950s and the preeminence of managerial values in the 1960s.

The methods of systems management in the 1950s similarly corresponded to the Cold War's ebb and flow. In the beginning, systems management concentrated on rapid development of high-performance systems. By the 1960s, however, cost and schedule concerns led to the creation of configuration management and phased planning. These two methods slowed development somewhat but in the process achieved greater precision in cost and schedule estimation. Systems management from 1965 onward embodied a mix of concerns, including scientists' bias for high performance, engineers' interest in reliability, and managers' need for cost prediction. It accurately reflected the shifting values of U.S. military and civilian organizations, leading to the rise and decline of scientists, engineers, and managers in technology development.

The Schriever Factor

Through two decades, from 1945 to 1965, when the Air Force developed its organization and processes for complex technology development, Bernard Schriever was at the center of events. He helped create SAB and ARDC in the 1940s. In the early 1950s in the Development Planning Office, Schriever helped establish systems analysis as a standard Air Force procedure to set requirements for new technologies. He headed the ballistic missile effort from 1953 to 1959, at which point he spearheaded the transition from ARDC to Air Force Systems Command, and the creation of the 375 series of systems management regulations for the Air Force. Finally, he was at the helm of AFSC during the early 1960s when Robert McNamara standardized systems management through the Department of Defense. Why was this one officer so influential in these events?

Part of the answer is fortune. It was Schriever's good luck to be well connected at the end of World War II. His relationship with Army Air Forces Chief Hap Arnold led to his assignment as scientific liaison immediately after the war. Schriever's career also encompassed a time when new technologies such as computers and missiles were ripe for development, and the Cold War provided convincing rationale for pouring massive funding into those programs.

Given these opportunities, Schriever's technically sound and politically astute efforts led to success. He made the most of the situation and was the right man at the right time. As he noted much later, the foundation for his later success was his belief that physical scientists like Theodore von Kármán and John von Neumann had the intelligence and foresight to peer into and shape the future. He believed that these scientists had greater vision than his military contemporaries and that he would be on sound footing if he took them seriously.

Farewell luncheon in June 1964 for ESD Commander Maj. Gen. Charles H. Terhune (far right). Joining him (left to right) are MITRE president C. W. Halligan, trustee James McCormack, and Vice Commander Brig. Gen. Otto J. Glasser. Terhune successfully negotiated with Halligan to remodel MITRE in the image of the Aerospace Corporation.

One consistent theme throughout Schriever's career was the periodic summoning of scientific and technical committees to evaluate technical and organizational issues. This was politically astute, considering the prestige of scientists in the late 1940s and early 1950s, and it was a worthwhile solicitation for their advice and vision. One early fruit of these connections was the Ridenour Report, produced while Schriever was one of Gen. Donald Putt's staff promoting the creation of a separate command for R&D.

After completing his stint as scientific liaison, Schriever worked with Ivan Getting in what became the development planning office. There he instituted development planning objectives that formalized the practice of RAND-style systems analysis to develop initial requirements for new Air Force systems. This replaced the former practice of simply asking operational officers what they wanted—a practice that generally resulted in incremental improvement of existing systems. He also helped create the weapon system concept through participation and diffusion of the Combat Ready Aircraft study. This resulted in the establishment of the Weapon System Project Offices that began the trend toward project management in the Air Force.

Military officers and historians most frequently remember Schriever for his path-clearing work in ballistic missiles. This began when Schriever first heard about the potential for small fusion weapons and he immediately realized they would make ballistic missiles feasible. He worked with famous mathematician John von Neumann to convince the Air Force and the Eisenhower administration that ICBMs were critical for defense. In this he was extraordinarily successful and he soon headed a separate development office in Inglewood. Even more amazingly, he and Trevor Gardner orchestrated a campaign to make ICBMs the nation's number-one-priority military development program. Using this authority and concerned with the slowdowns resulting from the Air Force's usual bureaucracy, Schriever and Gardner then pressed through the Gillette Procedures, a critical milestone in the development of R&D organization.

The Gillette Procedures gave Schriever unprecedented authority in technology development because they required ARDC, AMC, and others in the Air Force to aid him, without being able to slow him down. More important, the new procedures completely revamped the decision making and financial accounting for ICBMs. Instead of separate budgets and numerous organizations with which he had to coordinate, Schriever now had a single plan with a single budget and was responsible to a single authority, the Ballistic Missile Committee in the OSD. This streamlined process became the model for all major systems within a few years.

Upon Schriever's accession to be head of ARDC in 1959, Secretary of the Air Force James H. Douglas wanted him to fix the disorganization of electronic systems.[2] Schriever immediately began a series of internal and external studies to assess the situation. Despite concerns from MITRE and the Air Defense Systems Integration Division, which recommended moving command and control development to the OSD or NORAD, Schriever's Inglewood model became the basis for the subsequent reorganization of ARDC. At Inglewood, AMC complemented ARDC's Western Development Division with its Special Aircraft Projects Office, while R-W added its civilian technical experts. At Hanscom, Schriever folded ADSID into the new Air Force Command and Control Development Division, while AMC formed its counterpart Electronics Systems Center and MITRE was to be the expert civilian organization. Subsequently, Brig. Gen. Charles Terhune, who had been Schriever's technical deputy in Inglewood, took over to ensure the electronics group organized itself appropriately.

Although it initially rejected the concept as too radical, the Air Force adopted Schriever's recommended organization for systems management in early 1961, as the new McNamara regime offered the Air Force the DOD space mission if it reorganized space technologies and operations under a single command. By this time Schriever and Gen. Samuel Anderson had concluded that a single organization should handle the entire acquisition process instead of the awkward partnership between ARDC and AMC. Schriever separated research and logistics from acquisition, forming three separate organizations. The new Air Force Systems

Command under Schriever's control combined the old joint program offices of ARDC and AMC into integrated offices. The creation of the 375 systems management procedures standardized Inglewood processes across all four development divisions of AMC: Ballistic Systems, Space Systems, Aeronautical Systems, and Electronics Systems.

Finally, Schriever's Inglewood model, subsequently modified, became the DOD standard. McNamara found the 375 Procedures congenial to his centralizing goals, with the introduction of phased planning to ensure better cost and schedule estimates before program approval. In 1965 systems management became the DOD standard.

Schriever was critical to the Air Force because he was a strategic bridge between the technologists and the Air Force. He helped bring into the Air Force the three major methods created in the 1940s and 1950s to deal with complex technologies: operations research, project management, and systems engineering. Operations research was the scientists' means of analyzing complex human–machine systems. Project management became the manager's fundamental means of organizing complex systems. Systems engineering was the mechanism for coordinating the two endeavors. Schriever was the key mediator who transformed the methods into standard processes in the Air Force and the Department of Defense. Although Schriever and Terhune later lamented the formation of the 375 Procedures because they "removed flexibility," the fact remains that by using them the Air Force has been highly successful in developing and deploying technologies. When used as guidelines for technology development, the 375 Procedures encapsulated many of the critical lessons and methods developed in the 1950s. More than any other single individual, Bernard Schriever deserves the credit for merging scientific and engineering vision with military procedures to create the methods now standard throughout the Department of Defense.[3]

Paths Not Taken

While Schriever's Inglewood model swept through the Air Force between 1954 and 1962, those who worked on air defense and C^2 systems had developed an alternative model for technology development. The Hanscom model featured the development of prototypes through substantial interaction with end-user commands, and led logically to the integration of organizational structures at the highest level to ensure consistent strategy, tactics, and technology. The differences between the two models should not be exaggerated because they shared common methods of systems engineering and configuration management, but there were some significant variances. When the Inglewood model overtook Hanscom in the forming of AFCCDD and eventually the Electronics Systems Division, the Hanscom perspective became a minority viewpoint but it was not eradicated.

General Bergquist became a strong proponent of organizational integration as he pressed for the transfer of his ADSID from the Air Force to the OSD or to NORAD. His arguments and those of MITRE leaders like Robert Everett were logical but ultimately not persuasive. They argued that because air defense (and later the entire NORAD C^2 complex) integrated elements of the Army, Navy, and Air Force both in the United States and in Canada, the development organization creating technologies to automate command and control needed to have authority over all of those units.

It is ironic that Schriever persuasively argued for that same level of authority for ballistic missiles. The Air Force went along with his separate organization, understanding that the ballistic missiles so created eventually would be Air Force weapons. For ADSID and the SAGE system the organizational argument was even more relevant, but ultimately air defense and warning were not important enough to overcome service turf battles. Overall authority would not be forthcoming; rather, mere coordination would be the primary mechanism available to C^2 system developers.

As SAGE software development became the primary issue after 1955, Air Defense Command's importance grew because only it could supply the tactical doctrine and operational information required to build the SAGE master computer program. Unlike ballistic missiles, where tactics involved little more than pushing a button to launch, air defense required complex interactions among operators, pilots, missiles, and the various radar sites that made up the defense network. Thus ADC required a much larger up-front role in SAGE than Strategic Air Command had in ballistic missile development. Air defenses, which were clearly lower in the pecking order than bombers or ballistic missiles, required more interaction with the operational officers who dominated the Air Force's leadership, but the lack of priority and the need for interaction confused the Air Force's organization much more than did the relatively clear-cut "build and deliver" relationship that Schriever's organization had with SAC.

Later studies by RAND researchers clearly stated the problem of using the Inglewood model at ESD:

> Even with significant user participation, it has not been possible for the system developer to obtain a detailed description of user needs and operational requirements which could be translated into a coherent functional design and satisfactorily guide the system designer in the long-term development of command and control systems. Air Force experience with the development of the Strategic Air Command's Command and Control System (465-L), as well as with the development of the North American Air Defense Command's Command and Control System (425-L), has highlighted the difficulties of obtaining detailed descriptions of the current operations of the given

command and its functional requirements for command and control. Furthermore, even assuming the partially successful specification of the operational requirements for a command and control system (whether accomplished by the user, by some outside agency, or by an association of both), it is important to recognize that any such description is of necessity time-dated.[4]

At Hanscom, ESD and MITRE outwardly structured their efforts on the Inglewood model but locally they created mechanisms to ensure that their methods endured. With somewhat mixed results, ESD fostered intimate coordination with end-user organizations through the project offices. Without the authority to force cooperation with end-user commands such as SAC or Tactical Air Command, ESD and MITRE engineers and managers had to persuade commanders to work with them. Whereas the Inglewood model required up-front development of system specifications, at ESD the Air Force developed methods to create progressive deliveries of the C^2 systems. In essence, the first delivery was the prototype subsequently modified in later deliveries and augmentations of hardware and software. Configuration management became particularly critical in the deliveries of software.[5]

What if the Hanscom model instead of the Schriever model had prevailed? The fully implemented Hanscom model would have centralized C^2 development at the DOD level where McNamara could have controlled it. However, the model also emphasized intimate interactions with end-user commands and the development of prototypes for doing so. Would this have mitigated some of the later problems of pure paper studies that McNamara used so vigorously? We shall never know the answers to these questions but they remain important today. What if the Hanscom model of prototype development and developer–user interaction became the Air Force's standard? Perhaps for some technologies such as computing systems, explicit adoption of this approach might serve the military better.

Another point to mention is that systems engineering and technical direction (SE/TD) as implemented with the Air Force, Aerospace Corporation, and MITRE is not the only model possible. Although this method based on Schriever's belief in the utility of civilian experts is the dominant one in the Air Force, it is not the method used by the National Aeronautics and Space Administration, for example. NASA has believed in the need for in-house expertise and tries to keep about 10 percent of development work and funding internally so as to train its personnel. Even in the Air Force, not all officers believe in the efficacy of the SE/TD model.

A minority viewpoint within the Air Force has been that the Air Force should train and maintain its own technical expertise. One early proponent of this view was Col. Ed Hall in Schriever's own organization at Inglewood. Another was Col. Thomas Haig, who wrote in a mid-1960s assessment of the SE/TD concept that the split responsibilities and values inherent in the Air Force–nonprofit corporation relationships caused more problems than they solved. Haig, who had managed Air Force satellite programs in the early 1960s, was particularly irked by the pay differential between Air Force officers and their civilian counterparts at Aerospace and MITRE. Another irritant was the split reporting channels that violated the basic managerial and military principle of an undivided chain of command. Haig surveyed Air Force R&D officers and found that most of them registered complaints and wanted to limit the role of the civilians. With many more trained officers in the 1960s than in the 1950s, Haig saw no reason to continue the SE/TD relationship. He believed that Air Force officers had sufficient technical depth to take on the full responsibility, with perhaps some assistance from government in-house civilians.[6]

Indeed, one can ask as Haig did whether the use of civilians in powerful roles was an appropriate long-term course for the Air Force. Schriever justified the original arrangements on the basis of the Air Force's urgent need and the lack of technical officers in the early 1950s Air Force. The Air Force reduced this expertise shortfall by the middle to late 1960s through technical training at its own and other universities and the increased educational attainment of the United States population as a whole. It became much more feasible for Air Force technical officers to run R&D programs by the late 1960s, although the high turnover rate based on the Air Force's system of short-term tours in various duties was and remains a serious obstacle. The civilians in MITRE and Aerospace remain the repository of institutional memory for these technical programs. And, as Schriever himself believed, civilians brought a different vision to the Air Force than its own officers would have brought. To the present time, the system remains a civilian–military hybrid as Schriever originally established.

Founding the Future

Measured by the standard of military superiority, the Air Force's methods for developing new technologies have been quite effective. They likely have had a significant effect on the competitiveness of the private U.S. aerospace and computing industries as well. Those industries have been extraordinarily successful in international competition, the ultimate contemporary measure of business success or failure. Both industries developed systems engineering, project management, and configuration control in conjunction with the Air Force, and they employ those techniques as standards for organizing large-scale development. Research methods in both industries depend on a melange of funding agencies and concerns, and that leads to the generating of numerous ideas and technologies that can be integrated into new products using systems methods. The Air Force has been and remains one of the many providers of research funds.

Other studies of the military–industrial complex have come to rather pessimistic conclusions. Some found that the military distorted the course of science through its massive funding of specific disciplines of scientific interest. Others noted that scientists were willing partners and corrupted themselves in the process. Military direction of scientific, engineering, and economic processes also has been seen as a corruption of democracy. No doubt those criticisms are correct but the present work points out that military influence has a positive side as well, and not simply by virtue of the defeat of the Soviet Union.[7]

Generally speaking, the aerospace industry has been seen as a carrier of military bureaucracy and its supposed inability to innovate or react quickly to market conditions. The market for this industry, however, contains both military and civilian components. On the basis of international competition or perhaps even common sense, neither component can be classified as static and noninnovative. If the large aerospace companies have a problem it is that frequently they do not understand how to create mass-produced products for civilian purposes. Their products, however, typically consist of multimillion dollar aircraft or weapons and are not items that individuals typically buy. It is unreasonable to expect these companies, whose core competencies are so unlike those of consumer goods industries, to know how to operate under such different conditions.

The computer industry is one that markets its products to both the military and consumers, and its military background is unquestioned. Somehow the computer industry managed to have it both ways. If the management methods used in both industries are closely related, as has been shown here, then the differences between the two in terms of civilian utility resulted from the nature of their products rather than the nature of their management systems. The point to emphasize here is that both industries are extraordinarily innovative, and the management system used for both is one significant reason for their innovativeness.

What can be said without any doubt is that military and civilian innovators created organizational methods that became the standards for the Air Force and its contractors for new technology development. Until now it has been an unheralded achievement but one with far-reaching consequences. The United States defeated the Soviet Union largely on the basis of technological innovation and economic development. Part of the credit for U.S. technical and economic superiority must be given to the interaction of Air Force officers with civilian scientists, engineers, and managers. With the Cold War over and a new kind of competition based on commercial endeavor taking precedence, the full implications of this Cold War management system are yet to be seen. We are likely to find that many of the elements of systems management will continue to be important, but that others specific to the Cold War in which they were spawned will have to be given up or modified.

Notes

1. Interim rpt, von Kármán to H. H. Arnold, Gen of the AAF, "Where We Stand," Aug 22, 1945, in Michael H. Gorn, *Prophecy Fulfilled: "Toward New Horizons" and Its Legacy* (Washington, D.C.: AF Hist and Museums Prog, 1994), p 37.

2. Intvw, Bernard Schriever with author, Jul 6, 1999, USAF HSO.

3. See Stephen B. Johnson, "Three Approaches to Big Technology: Ops Research, Systems Engineering, and Project Management," *Technology and Culture* 38 (Oct 1997), pp 891–919.

4. S. M. Genensky and A. E. Wessel, "An Evolutionary Approach to AF Command and Control System Development," RAND Rpt RM-4178-PR, Jun 1964, pp 1–2.

5. See intvw, Charles Terhune with author, Oct 20, 1998, USAF HSO; intvw, Robert Everett with author, Oct 13, 1998, USAF HSO.

6. Lt Col Thomas O. Haig, "Systems Engineering and Technical Direction as a Management Concept," student research rpt no 67 (Washington, D.C.: Industrial College of the Armed Forces, 1966).

7. See the following scholarly studies: Stuart Leslie, *The Cold War and American Science* (New York: Columbia University Press, 1993); Walter A. McDougall, *The Heavens and the Earth: A Political History of the Space Age* (New York: Basic Books, 1985).

Glossary of Acronyms

AAS	American Astronautical Society
ADC	Air Defense Command
ADDC	Air Defense Direction Center
ADES	Air Defense Engineering Services
ADIS	Air Defense Integrated System
ADSEC	Air Defense Systems Engineering Committee
ADSID	Air Defense Systems Integration Division
ADSMO	Air Defense Systems Management Office
AF	Air Force
AFCCDD	Air Force Command and Control Development Division
AFCRC	Air Force Cambridge Research Center
AFCRL	Air Force Cambridge Research Laboratory
AFIT	Air Force Institute of Technology
AFLC	Air Force Logistics Command
AFR	Air Force Regulation
AFSC	Air Force Systems Command
AFSCM	Air Force Systems Command Manual
AIA	Aircraft Industries Association
AMC	Air Materiel Command
APL	Applied Physics Laboratory
ARDC	Air Research and Development Command
ARPA	Advanced Research Projects Agency
ASCA	Aircraft Stability and Control Analyzer
ASD	Aeronautical Systems Division
ATC	Air Training Command
AT&T	American Telephone and Telegraph
BMD	Ballistic Missile Division
BOMARC	Boeing-Michigan Aeronautical Research Center
BSD	Ballistic Systems Division
BTL	Bell Telephone Laboratories
C^2	Command and control
COC	Combat Operations Center
DCAS	Deputy Commander for Aerospace Systems

DCS/D   Deputy Chief of Staff, Development
DCS/M   Deputy Chief of Staff, Materiel
DCS/O   Deputy Chief of Staff, Operations
DCS/P   Deputy Chief of Staff, Personnel
DCS/S   Deputy Chief of Staff, Systems
DDRE   Director, Defense Research and Engineering
DEW   Distant Early Warning
DOD   Department of Defense
DPO   Development Planning Objective
ESC   Electronics Systems Center
ESD   Electronics Systems Division
ESSPO   Electronic Supporting System Project Office
FAA   Federal Aviation Administration
GMRD   Guided Missile Research Division (of Ramo-Wooldridge)
GOR   General operational requirement
HQ   Headquarters
IBM   International Business Machines
ICBM   Intercontinental ballistic missile
IRBM   Intermediate-range ballistic missile
JPL   Jet Propulsion Laboratory
MIT   Massachusetts Institute of Technology
NAA   North American Aviation
NACA   National Advisory Committee for Aeronautics
NASA   National Aeronautics and Space Administration
NCR   National Cash Register
NDRC   National Defense Research Committee
NORAD   North American Air Defense Command
ONR   Office of Naval Research
OR   Operations research
OSD   Office of the Secretary of Defense
OSRD   Office of Scientific Research and Development
PERT   Program Evaluation and Review Technique (or Research Task)
PPBS   Planning, Programming, and Budgeting System
RAND   Research and Development Corporation
R&D   Research and development
RCA   Radio Corporation of America
RDB   Research and Development Board
R-W   Ramo-Wooldridge Corporation
SAB   Scientific Advisory Board
SAC   Strategic Air Command
SAGE   Semi-Automatic Ground Environment
SAPO   Special Aircraft Projects Office
SDC   System Development Corporation

SDD	System Development Division
SE	Systems engineering
SecDef	Secretary of Defense
SRA	System Requirements Analysis
SRL	Systems Research Laboratory
SSD	Space Systems Division
STL	Space Technology Laboratories
TIR	Technical Information Release
TRW	Thompson-Ramo-Wooldridge
USAF	United States Air Force
WADC	Wright Air Development Center
WDD	Western Development Division
WE	Western Electric
WSPO	Weapon System Project Office

Glossary of Terms

Concurrency. In general, parallel processes or approaches. The word was coined by Bernard Schriever to describe his particular parallel approach to ballistic missile development.

Configuration Control. An organizational process to identify and control the hardware and software components and their interconnections within a design. By the mid-1960s it sometimes also included control over specifications. Under this process, changes to the design or components are not allowed unless they are approved by a configuration control board.

Configuration Management. An organizational process that expands configuration control by requiring all change requests to include cost and schedule estimates. Frequently, it also expands configuration control by having "paper traceability" of all materials and components from their manufacturing through operational use. This process was documented first in the AFSCM 375-1 regulations of the USAF in 1962.

Operations Research. A set of mathematical methods to analyze current human–machine operations in their organizational and operational environments. This research typically includes game theory, probability, and statistics.

Project Management. The set of management methods that organize tasks by end product. Usually the project is temporary and has an established completion date.

Systems Analysis. Operations research that is applied to future human–machine systems.

Systems Engineering. The set of methods used by engineers to coordinate complex technical development projects. Typically these methods include change control, analysis and documentation of specifications, and control of interfaces between technologies and organizations.

Systems Management. The set of organizational methods developed originally to create large-scale technologies in the aerospace and computing industries. It combines the methods of project management, systems engineering, systems analysis, and configuration management into a coherent management scheme. The term also refers to the 375-series systems management regulations created by the Air Force in 1960–61. Since the 1960s, it has spread beyond these industries.

Weapon System Concept. In the 1950s this term was used by the USAF to describe the procurement approach that planned for the entire life cycle of a weapon at the start of the weapon's development program. Thus the training, logistics, and testing for the weapon were planned for at the start, as were the design and production of the weapon itself.

Weapon System Project Office. The joint Air Research and Development Command–Air Materiel Command office that coordinated the development of a weapon system.

Notes on Sources

Research for this book took me to a number of archives around the United States, the most useful of which were the Air Force Historical Research Agency (AFHRA) at Maxwell AFB, in Alabama; the MITRE Archives in Bedford, Massachusetts; and the Lincoln Laboratory Archives in Lexington, Massachusetts.

For material about ballistic missiles, and for Air Force history in general, AFHRA is the best source. Many of the records formerly at the Space and Missile Center (SMC) in El Segundo, California, recently were shipped to AFHRA. Regrettably, not all of these were catalogued at AFHRA but they could be tracked from SMC's original shipping documents. The Aerospace Corporation, also in El Segundo, has a few records but most of the Aerospace records that are relevant to the present work are archived at TRW, and getting access to them is considerably more difficult. Fortunately, Davis Dyer's new history of TRW, *TRW: Pioneering Technology and Innovation Since 1900,* offered research on some of the organizations discussed in this study, as did John Lonnquest in his recent dissertation on Bernard Schriever. The Schriever papers and the Marvin C. Demler papers at AFHRA are excellent sources of organizational materials, as are ARDC and AMC command histories. Among the command histories, Ethel DeHaven's works were the most useful for my research.

For information about SAGE and air defense, the MITRE Archives unquestionably is the best source. MITRE has an extensive collection of materials from the SAGE period prior to the formation of MITRE and has collected substantial materials over the past decade. Lincoln Laboratory also has an extensive collection on SAGE. The Hanscom Research Library, Hanscom History Office, and AFHRA have command histories for the succession of Air Force organizations that dealt with air defense in the 1950s and early 1960s. Kent Redmond and Thomas Smith's 1997 manuscript, *From Whirlwind to MITRE: The R&D Story of the SAGE Air Defense Computer,* a copy of which is filed at MITRE, was an invaluable source, as was their earlier history of Whirlwind. Atsushi Akera's recent dissertation also includes useful Whirlwind material. John Jacobs' published "personal history" of SAGE also held the clues to many of the events and much of the politics involving SAGE, particularly its all-critical systems office. SAGE also has received recent attention from historians Tom Hughes and Paul Edwards and journalist Robert Buderi. Air Force Space Command has significant holdings related to air defense, but those are not catalogued. They can be accessed only through the Air Force Space Command History Office (AFSpC) staff, who are very helpful but also very busy. Many of the records remain classified so they are less useful to outside historians without security clearances. Finally, the System Development Corporation files in the Burroughs Collection housed at the Charles Babbage Institute, University of Minnesota, include papers on RAND and SDC, as do the Herbert Simon and Allen Newell papers at the Carnegie Mellon University Archives.

Archives Listing

Aerospace   Aerospace Corporation Archives, El Segundo, California

AFHRA   Air Force Historical Research Agency, Maxwell AFB, Montgomery, Alabama

AFSpC   Air Force Space Command History Office, Colorado Springs, Colorado

CBI   Charles Babbage Institute, University of Minnesota, Minneapolis, Minnesota

CMU   Carnegie Mellon University Archives, Pittsburgh, Pennsylvania

HAEUI   Historical Archives of the European University Institute

Hanscom   Hanscom AFB History Office, Lexington, Massachusetts

HRL   Hanscom AFB Research Library, Massachusetts Institute of Technology, Lexington, Massachusetts

JPLA   Jet Propulsion Laboratory Archives, Pasadena, California

Lincoln   Lincoln Laboratory Archives, Lexington, Massachusetts

LOC/SPP   Library of Congress, Washington, D.C., Samuel Phillips Papers

MDP   Marvin C. Demler Papers

MIT   Massachusetts Institute of Technology, Lexington, Massachusetts

MITRE   MITRE Archives, Bedford, Massachusetts

SDCP   System Development Corporation Papers

SMC   United States Air Force Space and Missile Center, El Segundo, California

USAF HSO   United States Air Force History Support Office, Bolling AFB, Washington, D.C.

Bibliography

Primary Sources

Books

Ackoff, Russell L., and Patrick Rivett. *A Manager's Guide to Operations Research.* New York: John Wiley and Sons, 1963.

Barish, Norman N. *Systems Analysis for Effective Administration.* New York: Funk and Wagnalls, 1951.

Barnard, Chester I. *Functions of the Executive.* Cambridge: Harvard University Press, 1938.

Churchman, C. West, Russell L. Ackoff, and E. Leonard Arnoff. *Introduction to Operations Research.* New York: John Wiley and Sons, 1957.

Cleland, David I., and William R. King. *Systems Analysis and Project Management.* New York: McGraw-Hill, 1968.

Davis, Ralph Currier. *Industrial Organization and Management.* 3d ed. New York: Harper and Brothers, 1957.

Eckman, Donald P., ed. *Systems: Research and Design, Proceedings of the First Systems Symposium at Case Institute of Technology.* New York: John Wiley and Sons, 1961.

Flagle, Charles D., William H. Huggins, and Robert H. Roy. *Operations Research and Systems Engineering.* Baltimore: Johns Hopkins University Press, 1960.

Goode, Harry H., and Robert E. Machol. *Systems Engineering.* New York: McGraw-Hill, 1957.

Groves, Lt. Gen. Leslie R. "The A-Bomb Program." In *Science, Technology, and Management,* edited by Fremont Kast and James Rosenzweig, 33–40. New York: McGraw-Hill, 1963.

Hall, Arthur D. *A Methodology for Systems Engineering.* Princeton, N.J.: D. Van Nostrand, 1962.

Hitch, Charles J. *The Economics of Defense in the Nuclear Age.* Cambridge: Harvard University Press, 1960.

———. *Decision Making for Defense.* Berkeley: University of California Press, 1967.

Johnson, Ellis. "Operations Research in the World Crisis in Science and Technology." In *Operations Research and Systems Engineering,* edited by Charles D. Flagle, William H. Huggins, and Robert H. Roy, 28–57. Baltimore: Johns Hopkins University Press, 1960.

Kast, Fremont, and James Rosenzweig, eds. *Science, Technology, and Management.* New York: McGraw-Hill, 1963.

MacColl, L. A. *Fundamental Theory of Servomechanisms.* New York: D. Van Nostrand, 1946.

Machol, Robert E., Wilson P. Tanner, Jr., and Samuel N. Alexander. *System Engineering Handbook.* New York: McGraw-Hill, 1965.

Martino, Rocco. *Project Management and Control.* Vol. 1, *Finding the Critical Path.* New York: American Management Association, 1964.

————. *Project Management.* Wayne, Penn.: MDI Publications, 1968.

McCloskey, Joseph, and Florence N. Trefethen, eds. *Operations Research for Management.* Baltimore: Johns Hopkins University Press, 1954.

McKean, Roland N. *Efficiency in Government Through Systems Analysis, with Emphasis on Water Resources Development, A RAND Corporation Research Study.* New York: John Wiley and Sons, 1958.

McNamara, Robert. *The Essence of Security.* New York: Harper and Row, 1968.

Morse, Philip M., and George F. Kimball. *Methods of Operations Research.* New York: John Wiley and Sons, 1951.

Novick, D. *Program Budgeting.* Cambridge: Harvard University Press, 1965.

Peck, Merton J., and Frederic M. Scherer. *The Weapons Acquisition Process: An Economic Analysis.* Boston: Division of Research, Graduate School of Business Administration, Harvard University, 1962.

Proceedings, National Symposium on Quality Control and Reliability in Electronics. Sponsored by the Professional Group on Quality Control, Institute of Radio Engineers and the Electronics Technical Committee, American Society for Quality Control. New York City, November 12–13, 1954. New York: Institute of Radio Engineers, 1955.

Raborn, Vice Adm. W. F., Jr. "Management of the Navy's Fleet Ballistic Missile Program." In *Science, Technology, and Management,* edited by Fremont Kast and James Rosenzweig, 148. New York: McGraw-Hill, 1963.

Ries, John C. *Management of Defense: Organization and Control of the United States Armed Services.* Baltimore: Johns Hopkins University Press, 1964.

Roethlisberger, F. J., and William J. Dickson. *Management and the Worker.* Cambridge: Harvard University Press, 1939.

Ross, H. John. *Technique of Systems and Procedures.* New York: Office Research Institute, 1948.

Roy, Robert H. *The Administrative Process.* Baltimore: Johns Hopkins University Press, 1958.

————. "The Development and Future of Operations Research and Systems Engineering." In *Operations Research and Systems Engineering,* edited by Charles D. Flagle, William H. Huggins, and Robert H. Roy, 8–27. Baltimore: Johns Hopkins University Press, 1960.

Shannon, Claude E., and Warren Weaver. *The Mathematical Theory of Communication.* Urbana: University of Illinois Press, 1949.

Simon, Herbert A. *Administrative Behavior: A Study of Decision-Making Processes in Administrative Organization.* New York: Macmillan, 1946.

————. *Models of Man, Social and Rational: Mathematical Essays on Rational Human Behavior in a Social Setting.* New York: John Wiley and Sons, 1957.

Sutton, George Paul. *Rocket Propulsion Elements: An Introduction to the Engineering of Rockets.* 6th ed. New York: John Wiley and Sons, 1992.

Taylor, Frederick W. *Shop Management.* New York: American Society of Mechanical Engineers, 1903, and New York: Harper and Brothers, 1911).

———. *The Principles of Scientific Management.* New York: Harper and Brothers, 1911.

Weber, Max. *The Theory of Social and Economic Organization.* Translated by A. M. Henderson and Talcott Parsons. New York: Free Press, 1947.

Wiener, Norbert. *Cybernetics: or Control and Communication in the Animal and the Machine.* Cambridge: MIT Press, 1965.

Reports and Government Documents

Army Air Forces Scientific Advisory Group. "Toward New Horizons." Report to General H. H. Arnold, December 15, 1945. In *Prophecy Fulfilled: "Toward New Horizons" and Its Legacy,* edited by Michael H. Gorn. Washington, D.C.: Air Force History and Museums Program, 1994.

Ballistic Missile Organization History Office Staff. *Chronology of the Ballistic Missile Organization: 1945–1990.* 1993.

Bellis, Benjamin N. "The Requirements for Configuration Management During Concurrency." Paper presented at the Air Force Systems Command Management Conference, Monterey, Calif., May 2–4, 1962. Air Force Historical Research Agency, Maxwell AFB, Montgomery, Ala., microfilm 26254, pp. 2-24-1–8.

Biel, W. C., R. L. Chapman, J. L. Kennedy, and A. Newell. "The Systems Research Laboratory's Air Defense Experiments." Report P-1202. Santa Monica, Calif.: RAND Corporation, 1957.

Black, Maj. William M., and Capt. Kemper W. Baker. *History of the Directorate of Research and Development, Office Deputy Chief of Staff, Development, HQ USAF for Period 1 July 1950–31 December 1950.* Air Force Historical Research Agency, Maxwell AFB, Montgomery, Ala., K140.01.

Bush, Vannevar. *Science, the Endless Frontier.* Washington, D.C.: U.S. Government Printing Office, 1945.

CECLES—ELDO 1960–1965. Report to the Council of Europe, Paris, December 23, 1965. Historical Archives of the European University Institute.

DeHaven, Ethel M. *Aerospace: The Evolution of USAF Weapons Acquisition Policy 1945–1961.* Vol. 6, *History of DCAS—1961.* Los Angeles: Air Force Systems Command, SMC History Office, 1962.

———. *AMC Participation in the AF Missiles Program through December 1957.* Inglewood, Calif: Office of Information Services, AMC Ballistic Missile Office, 1957.

Department of the Navy Special Projects Office. "Program Evaluation Research Task, Summary Report, Phase 1." Washington, D.C.: Bureau of Naval Weapons, Department of the Navy, 1958.

Executive Office of the President, Bureau of the Budget. *Report to the President on Government Contracting for Research and Development.* Bell Report. April 30, 1962. In *Exploring the Unknown: Selected Documents in the History of the U. S. Civil Space Program,* edited by John M. Logsdon, SP-4218, 652–72. Washington, D.C.: National Aeronautics and Space Administration, 1996.

Genensky, S. M., and A. E. Wessel. "An Evolutionary Approach to Air Force Command and Control System Development." Report RM-4178-PR, June 1964. MITRE Archives.

Jordan, Harry C. *History of Ballistic Systems Division, Air Force Systems Command, The Division Pronaos.* Norton AFB, Calif.: Historical Division, Office of Information, Ballistic Systems Division, Air Force Systems Command, April 3, 1963. MITRE Archives, box 1753.

Kennedy, J. L. "RAND Extension of Bavelas Project." In "Report of a Seminar on Organizational Science," edited by M. M. Flood. RAND research memorandum RM-709, October 29, 1951. Carnegie Mellon University Archives, Herbert Simon Papers.

Littlejohn, Edward, and C. J. McClain. *The Accounting Machine Industry, A Study in Competition.* Detroit: Burroughs Adding Machine Company, June 1950. Charles Babbage Institute, University of Minnesota, Burroughs Collection, box 1, Corporate Product Planning 1950–1964; folder, The Accounting Machine Industry 1950.

Livingston, J. Sterling. "Advanced Techniques for Program/Funds Management." Paper presented at the Air Force Systems Command Management Conference, Monterey, Calif., May 2–4, 1962. Air Force Historical Research Agency, Maxwell AFB, Montgomery, Ala., microfilm 26254, paper no. 4-11.

Luke, Maj. Ernst P., and Ernst Lange. "Weapon System Evaluation and Reliability." Paper presented at the Symposium on Guided Missile Reliability, November 2–4, 1955. Sponsored by the Department of Defense, under auspices of Air Research and Development Command, AF-WP-O-21. Air Force Historical Research Agency, Maxwell AFB, Montgomery, Ala., microfilm 26254.

Lulejian, Col. Norair M. "Scheduling Invention." Paper presented at the Air Force Systems Command Management Conference, Monterey, Calif., May 2–4, 1962. Air Force Historical Research Agency, Maxwell AFB, Montgomery, Ala., microfilm 26254, paper no. 1-4.

Maddox, Lt. Col. Robert A., and Lt. Col. John L. Gregory, Jr. *History of the Directorate of Research and Development Office, Deputy Chief of Staff, Development, Headquarters, USAF, 1 July 1951 to 31 December 1951.* Air Force Historical Research Agency, Maxwell AFB, Montgomery, Ala., K140.01.

—————. *History of the Directorate of Research and Development Office, Deputy Chief of Staff, Development, Headquarters USAF, 1 January 1952 through 30 June 1952.* Air Force Historical Research Agency, Maxwell AFB, Montgomery, Ala., K140.01.

Malina, Frank J. "Development and Flight Performance of a High Altitude Sounding Rocket the 'WAC Corporal.'" Jet Propulsion Laboratory report no. 4-18, January 24, 1946. Jet Propulsion Laboratory Archives, historian index no. 03 00066 XF.

Marshall, A. W., and W. H. Meckling. "Predictability of the Cost, Time and Success of Development." Report P-1821. Santa Monica, Calif.: RAND Corporation, 1959.

McMullen, Richard F. *The Birth of SAGE 1951–1958.* ADC historical study no. 33, MIT Lincoln Archives, accession #358587.

Murphy, Howard R. *History of the Air Force Command and Control Development Division, 16 November 1959–31 March 1961.* Vol. 1, *Prologue to the Hanscom Complex.* Air Force Systems Command historical publication 64-32-I. Bedford,

Mass.: Historical Division, Office of Information, Electronic Systems Division, Hanscom Field, Air Force Systems Command, 1964.

Office of History, Headquarters, Space Division. *Space and Missile Systems Organization: A Chronology, 1954–1979.* USAF Space and Missile Center Archives.

Perry, Robert L. *Origins of the USAF Space Program 1945–1956.* Originally printed as Volume 5, *History of the Deputy Commander (AFSC) for Aerospace Systems 1961.* Los Angeles: Space Systems Division, Air Force Space Command. Space and Missile Systems Center History Office, 1997.

Pickering, William. "Management Techniques for the Management and Development of Weapon Systems." Address presented at the Industrial College of the Armed Forces, Washington, D.C., January 22, 1958. Jet Propulsion Laboratory Archives, document 3-655.

Proceedings of the First International Conference on Operational Research. Baltimore: Operations Research Society of America, 1957.

Proceedings of the Symposium on Guided Missile Reliability. November 1955. Sponsored by the Department of Defense, under auspices of Air Research and Development Command, AF-WP-O-21, September 1956, 56RDZ-12531, Wright-Patterson AFB, Ohio. Air Force Historical Research Agency, Maxwell AFB, Montgomery, Ala., microfilm 26254.

Redmond, Kent C., and Harry C. Jordan. *Air Defense Management, 1950–1960: The Air Defense Systems Integration Division.* Bedford, Mass.: Historical Branch, Office of Information, Air Force Command and Control Development Division, Laurence G. Hanscom Field, 1961.

Roberts, Nicholas. *History of the Directorate of Requirements, Deputy Chief of Staff, Development, 1 July 1949–30 June 1950.* Air Force Historical Research Agency, Maxwell AFB, Montgomery, Ala., K140.01.

Robillard, G. *The Explorer Rocket Research Program.* JPL publication no. 145. October 31, 1958, Jet Propulsion Laboratory Archives, document R093.

Scientific Advisory Board. *Research and Development in the United States Air Force.* Ridenour Report. September 21, 1949. Air Force Historical Research Agency, Maxwell AFB, Montgomery, Ala., 168.1511-1.

Semi-Annual Historical Report for the Deputy Chief of Staff, Development, 1 January 1953–30 June 1953. Air Force Historical Research Agency, Maxwell AFB, Montgomery, Ala., K140.01.

Semi-Annual Historical Report for the Deputy Chief of Staff, Development, Vol. 1, 1 January 1954–30 June 1954. Air Force Historical Research Agency, Maxwell AFB, Montgomery, Ala., K140.01.

Serig, H. W. *Project Lincoln Case History.* Bedford, Mass.: Air Force Cambridge Research Center, 1952. MIT Lincoln Laboratory Archives, U-13.050.

System Development Corporation. Santa Monica, Calif.: System Development Corporation, 1957–1959. Charles Babbage Institute, University of Minnesota, 90, Burroughs Collection, System Development Corporation series, box 1, "SDC Descriptive Booklets" folder.

System Requirements Analysis Orientation Guide. San Bernardino, Calif.: Space Technology Laboratories, Inc., Thompson-Ramo-Wooldridge, Inc., for the Minuteman Program Office of U.S. Air Force Systems Command, ca. 1960. Library of Congress, Samuel Phillips Papers, box 18.

System Training Program Staff. "The System Training Program for the Air Defense Command (U)." Report RM-1157-AD. Santa Monica, Calif.: RAND Corporation, 1953.

Thornton, J. P. *History of the Deputy Chief of Staff, Development, 1 July to 31 December 1950.* Air Force Historical Research Agency, Maxwell AFB, Montgomery, Ala., K140.01.

————. *History of the Deputy Chief of Staff/Development, 1 January to 30 June 1951.* Air Force Historical Research Agency, Maxwell AFB, Montgomery, Ala., K140.01.

————. *History of the Deputy Chief of Staff, Development, July 1951 to June 1952.* Air Force Historical Research Agency, Maxwell AFB, Montgomery, Ala., K140.01.

Toward the Endless Frontier: History of the Committee on Science and Technology, 1959–1979. Washington, D.C.: U.S. Government Printing Office, 1980.

United Kingdom Air Ministry. *The Origins and Development of Operational Research in the Royal Air Force.* Publication 3368. London: Her Majesty's Stationery Office, 1963.

U.S. Air Force. *Organization Charts, Headquarters USAF, 1947–1984. AFP-210-5.* Washington, D.C.: Office of Air Force History, 1984.

U.S. Air Force. Viewgraphs for Air Force Institute of Technology system program management course, probably presented by Samuel C. Phillips, ca. 1963. Library of Congress, Samuel Phillips Papers, box 37.

U.S. Air Force Ballistic Missile Division (HQ ARDC). "AFBMD Management and Organization for Ballistic Missile Weapon Systems Development," WD-57-05415, December 1957. El Segundo, Calif.: Space and Missile Center History Office Archives.

U.S. Air Force Deputy Chief of Staff, Development, Headquarters USAF. "Combat Ready Aircraft: How Better Management Can Improve the Combat Readiness of the Air Force." Special report based on an Air Force study completed April 1951. Air Force Historical Research Agency, Maxwell AFB, Montgomery, Ala., Marvin C. Demler Papers, 168.7265-236.

U.S. Air Force, Headquarters, Air Force Systems Command. *AFSC Management Objectives, FY 1964.* August 1963. Library of Congress, Samuel Phillips Papers, box 18.

U.S. Air Force, Headquarters, Air Force Systems Command. *Air Force Systems Command Manual, Systems Management, AFSCM 375-1: Configuration Management During the Acquisition Phase.* June 1, 1962. Library of Congress, Samuel Phillips Papers, box 18.

U.S. Air Force, Headquarters, Air Force Systems Command. *Program Management Instruction No. 1-7: Systems Management Procedures for Designated Systems.* Draft dated August 3, 1961. Library of Congress, Samuel Phillips Papers, box 18.

U.S. Air Force, Headquarters, Air Force Systems Command. *Rainbow Reporting: Systems Data Presentation and Reporting Procedures, Program Management Instruction 1-5.* January 15, 1963. Library of Congress, Samuel Phillips Papers, box 18.

U.S. Air Force, Headquarters, Air Force Systems Command. *A Summary of Lessons Learned from Air Force Management Surveys.* AFSCP 375-2, June 1, 1963. Library of Congress, Samuel Phillips Papers, box 36.

U.S. Air Force, Headquarters, Air Force Systems Command, Ballistic Systems Division, Minuteman Program. "Discussion of Systems Engineering," Viewgraphs, perhaps for class at Wright Air Development Center, ca. 1964. Library of Congress, Samuel Phillips Papers, box 18.

U.S. Air Force, Headquarters, Air Force Systems Command, Space Systems Division. *Program 624A Management Philosophy and Technical Approach.* June 9, 1962. Library of Congress, Samuel Phillips Papers, box 18.

U.S. Air Force Systems Command Management Conference Proceedings, Monterey, Calif., May 2–4, 1962. Air Force Historical Research Agency, Maxwell AFB, Montgomery, Ala., microfilm 26254.

U.S. Department of the Air Force. *Air Force Regulation No. 375-1, Systems Management, Management of Systems Programs.* February 12, 1962. Library of Congress, Samuel Phillips Papers, box 37.

U.S. Department of the Air Force. *Air Force Regulation No. 375-2, Systems Management, System Program Office.* February 12, 1962. Library of Congress, Samuel Phillips Papers, box 37.

U.S. Department of the Air Force. *Air Force Regulation No. 375-3, Systems Management, System Program Director.* February 12, 1962. Library of Congress, Samuel Phillips Papers, box 37.

U.S. Department of the Air Force. *Air Force Regulation No. 375-4, Systems Management, System Program Documentation.* February 12, 1962. Library of Congress, Samuel Phillips Papers, box 37.

U.S. Department of the Air Force. *Air Force Regulation No. 80-5, Research and Development, Reliability Program for Systems, Subsystems, and Equipments.* June 4, 1962. Library of Congress, Samuel Phillips Papers, box 37.

U.S. Department of Defense. Directive 3200.9. "Initiation of Engineering and Operational System Development." July 1, 1965.

U.S. House. Committee on Armed Services. *Hearings, Weapons System Management and Team System Concept in Government Contracting.* 86th Cong., 1st sess. Washington, D.C.: U.S. Government Printing Office, 1959.

U.S. House. Committee on Armed Services, Subcommittee for Special Investigations. *Weapons System Management and Team System Concept in Government Contracting.* 86th Cong., 1st sess. Washington, D.C.: U.S. Government Printing Office, 1959.

U.S. House. Committee on Government Operations. *Organization and Management of Missile Programs.* 86th Cong., 1st sess., report no. 1121. Washington, D.C.: U.S. Government Printing Office, 1959.

U.S. House. Committee on Government Operations. *Hearings, Organization and Management of Missile Programs.* 86th Cong., 1st sess. Washington, D.C.: U.S. Government Printing Office, 1959.

U.S. House. Committee on Government Operations. *Air Force Ballistic Missile Management: Formation of Aerospace Corporation.* 87th Cong., 1st sess., report no. 324. Washington, D.C.: U.S. Government Printing Office, 1961.

U.S. Senate. Committee on Appropriations, Chavez Subcommittee. *Hearings on Department of Defense Appropriations for 1960.* 86th Cong. 1st sess. Washington, D.C.: U.S. Government Printing Office, 1960.

Value Engineering Methods Manual. Seattle: Boeing Company, Aero-Space Division, D2-34567, ca. 1962. Library of Congress, Samuel Phillips Papers, box 19.

Von Kármán, Theodore. "Where We Stand." Interim Report to H. H. Arnold, General of the Army Air Forces, August 22, 1945. In *Prophecy Fulfilled: "Toward New Horizons" and Its Legacy,* edited by Michael H. Gorn. Washington, D.C.: Air Force History and Museums Program, 1994.

"Winter Study Report: The Challenge of Command and Control." March 31, 1961. MITRE Archives.

Articles

Bergen, William B. "New Management Approach at Martin." *Aviation Age* 20, no. 6 (June 1954) 39–47.

Bode, Hendrick. "The Systems Approach." A report to the Committee on Science and Astronautics, U.S. House of Representatives, by the National Academy of Science. *Applied Science and Technological Progress* (June 1967): 73–94.

Boulding, Kenneth E. "General Systems Theory—The Skeleton of Science." *Management Science* 2, no. 3 (April 1956): 197–208.

Davis, Keith. "The Role of Project Management in Scientific Manufacturing." *IRE Transactions on Engineering Management* EM-9, no. 3 (September 1962): 109–13.

Forrester, J. W. "Industrial Dynamics—a Major Breakthrough for Decision Makers." *Harvard Business Review* 36 (July–August 1958): 37–66.

Gates, C. R. "Systems Analysis." In *Seminar Proceedings, Systems Engineering in Space Exploration, May 1–June 5, 1963.* Jet Propulsion Laboratory, California Institute of Technology, Pasadena, June 1, 1965. Jet Propulsion Laboratory Archives, historian index no. 02 00221 XF.

Hopkins, R. C. "A Systematic Procedure for System Development." *IRE Transactions on Engineering Management* EM-8, no. 2 (June 1961): 77–86.

Johnson, Samuel C., and Conrad Jones. "How to Organize for New Products." *Harvard Business Review* 35, no. 3 (May–June 1957): 49–62.

Katz, Abraham. "An Industrial Dynamic Approach to the Management of Research and Development." *IRE Transactions on Engineering Management* EM-6, no. 3 (September 1959): 75–80.

Lanier, H. F. "Organizing for Large Engineering Projects." *Machine Design* 28 (December 27, 1956): 54–60.

Lewis, Howard T., and Charles A. Livesey. "Materials Management, A Problem of the Airframe Industry." *Business Research Studies* 31, no. 3 (July 1944).

Loosbrock, John F. "The USAF Ballistic Missile Program." *Air Force* 41, no. 3 (March 1958): 84–95.

Malcolm, D. G., J. H. Roseboom, C. E. Clark, and W. Fazar. "Applications of a Technique for R&D Program Evaluation (PERT)." *Operations Research* 7, no. 5 (1959): 646–69.

McLaughlin, John J. "The Air Force's Management Revolution." *Air Force and Space Digest* 45, no. 9 (September 1962): 98–106.

Nyquist, H. "Certain Topics in Telegraph Transmission Theory." *A.I.E.E. Transactions* 47 (April 1928): 627.

Putt, Maj. Gen. Donald C. "Freedom from Want and Waste." *Air Force* 35, no. 1 (1952): 28–32, 43.

"RDB—The Planners." *Aviation Age* 19, no. 6 (June 1953): 30–31.

Reed, Robert J. "New AF Policy Means More Competition—More Selling." *Aviation Age* 19, no. 8 (August 1953): 20–23.

Rubenstein, Albert H. "Setting Criteria for R&D." *Harvard Business Review* 35, no. 1 (January–February 1957): 95–104.

Simon, Herbert A. "Application of Servomechanism Theory to Production Control." *Econometrica* 20 (April 1952): 247–68.

Uhl, Edward G. "Applying the Systems Method to Air Weapons Development." *Aviation Age* 20, no. 2 (February 1954): 20–23.

Wilson, Randle C. "Problems of R&D Management." *Harvard Business Review* 37, no. 1 (January–February 1959): 128–36.

Zappacosta, Amedeo D. "Value Engineering." *IRE Transactions on Product Engineering and Production* PEP-6, no. 1 (March 1962): 4–10. Library of Congress, Samuel Phillips Papers, box 19.

Interviews

Everett, Robert. Interview by the author, tape recording, Grand Forks, N.D. 1 and 13 October 1998. USAF Headquarters History Support Office, Washington, D.C.

Getting, Ivan. Interview by the author, tape recording, Grand Forks, N.D. 30 October and 6 November 1998. USAF Headquarters History Support Office, Washington, D.C.

Haig, Tom. Interview by the author. 8 October 1996.

Hall, Col. Edward N. Interview by Jack Neufeld, USAF Oral History Program. 11 July 1989. Air Force Historical Research Agency, Maxwell AFB, Montgomery, Ala., K239.0512-1820.

Morriss, Ben. Interview by the author, tape recording, Grand Forks, N.D. 25 September 1998. USAF Headquarters History Support Office, Washington, D.C.

Phillips, Lt. Gen. Samuel. Interview by Tom Ray. 22 July 1970. NASA Headquarters History Office, "Phillips, Samuel C., Interviews" folder 001701, LEK 1/13/4.

Phillips, Maj. Gen. James F. Interviewer unknown. "Historical Documentation." USAF Oral History Program. 24 June 1966. Air Force Historical Research Agency, Maxwell AFB, Montgomery, Ala., K239.0512-787.

Putt, Lt. Gen. Donald L. Interview by James C. Hasdorff, USAF Oral History Program. 1–3 April 1974. Air Force Historical Research Agency, Maxwell AFB, Montgomery, Ala., K239.0512-724.

Schriever, Bernard. Interview by the author, tape recording, Grand Forks, N.D. 4 and 25 March, 13 April, 16 June, and 6 July 1999. USAF Headquarters History Support Office, Washington, D.C.

The USAF and the Culture of Innovation

Schriever, Gen. Bernard A. Interview by Dr. Edgar F. Puryear, Jr., USAF Oral History Program. 15 and 29 June 1977. Air Force Historical Research Agency, Maxwell AFB, Montgomery, Ala., K239.0512-1492.

Soper, Col. Ray E. Interview notes transcribed on the final day of his service with USAF before retirement. 29 November 1966. BSD(BEH) NAFB, Cal 92409, Air Force Historical Research Agency, Maxwell AFB, Montgomery, Ala., microfilm 30015.

Terhune, Charles. Interview by author, tape recording, Grand Forks, N.D. 24 and 30 September, 14 and 20 October 1998. USAF Headquarters History Support Office, Washington, D.C.

Secondary Works

Books

Abele, Ray. *The Burroughs Story.* First draft. Detroit: Burroughs Corporation, 1975. CBI, Burroughs Collection, box 2: Histories of Burroughs, 1886–1986.

Aitken, Hugh G. H. *Taylorism at Watertown Arsenal: Scientific Management in Action, 1908–1915.* Cambridge: Harvard University Press, 1960.

Armacost, Michael H. *The Politics of Weapons Innovation: The Thor–Jupiter Controversy.* New York: Columbia University Press, 1969.

Armytage, W. H. G. *The Rise of the Technocrats: A Social History.* London: Routledge and Kegan Paul, 1965.

Art, Robert J. *The TFX Decision: McNamara and the Military.* Boston: Little, Brown and Company, 1968.

Baum, Claude. *The System Builders: The Story of SDC.* Santa Monica, Calif.: System Development Corporation, 1981.

Beard, Edmund. *Developing the ICBM: A Study in Bureaucratic Politics.* New York: Columbia University Press, 1976.

Beniger, James R. *The Control Revolution: Technological and Economic Origins of the Information Society.* Cambridge: Harvard University Press, 1986.

Benson, Lawrence R. *Acquisition Management in the United States Air Force and Its Predecessors.* Washington, D.C.: Air Force History and Museums Program, 1997.

Beyerchen, Alan. "From Radio to Radar: Interwar Military Adaptation to Technological Change in Germany, the United Kingdom, and the United States." In *Military Innovation in the Interwar Period,* edited by Williamson Murray and Allan R. Millett, 265–99. Cambridge: Cambridge University Press, 1996.

Bijker, Wiebe E., Thomas P. Hughes, and Trevor Pinch, eds. *The Social Construction of Technological Systems.* Cambridge: MIT Press, 1987.

Bilstein, Roger E. *Stages to Saturn: A Technological History of the Apollo/Saturn Launch Vehicles.* Document SP-4206. Washington, D.C.: National Aeronautics and Space Administration, 1980.

———. *Flight in America, From the Wrights to the Astronauts.* Rev. ed. Baltimore: Johns Hopkins University Press, 1994.

———. *The American Aerospace Industry.* New York: Twayne Publishers, 1996.

Borklund, C. W. *Men of the Pentagon from Forrestal to McNamara.* New York: Frederick A. Praeger, 1966.

————. *The Department of Defense.* New York: Frederick A. Praeger, 1968.

Bright, Charles D., ed. *Historical Dictionary of the U.S. Air Force.* New York: Greenwood Press, 1992.

Bromberg, Joan Lisa. *NASA and the Space Industry.* Baltimore: Johns Hopkins University Press, 1999.

Brown, Michael E. *Flying Blind: The Politics of the U.S. Strategic Bomber Program.* Ithaca, N.Y.: Cornell University Press, 1992.

Buderi, Robert. *The Invention That Changed the World.* New York: Simon and Schuster, 1996.

Bulkeley, Rip. *The Sputnik Crisis and Early United States Space Policy.* Bloomington: Indiana University Press, 1991.

Campbell-Kelly, Martin, and William Aspray. *Computer: A History of the Information Machine.* New York: Basic Books, 1996.

Chandler, Alfred D., Jr. *Strategy and Structure, Chapters in the History of Industrial Enterprise.* Cambridge: MIT Press, 1962.

————. *The Visible Hand: The Managerial Revolution in American Business.* Cambridge: Harvard University Press, Belknap Press, 1977.

————. *Scale and Scope: The Dynamics of Industrial Capitalism.* Cambridge: Harvard University Press, Belknap Press, 1990.

Chapman, Richard L. *Project Management in NASA, the System and the Men.* Document SP-324. Washington, D.C.: National Aeronautics and Space Administration, 1973.

Constant, Edward W. III. *The Origins of the Turbojet Revolution.* Baltimore: Johns Hopkins University Press, 1980.

Cortada, James. *Before the Computer: IBM, NCR, Burroughs, Remington Rand and the Industry They Created, 1865–1956.* Princeton, N.J.: Princeton University Press, 1993.

Crockatt, Richard. *The Fifty Years War: The United States and the Soviet Union in World Politics, 1941–1991.* New York: Routledge, 1995.

Davidson, Frank P., and C. Lawrence Meador, eds. *Macro-Engineering and the Future, A Management Perspective.* Boulder, Colo.: Westview Press, 1982.

Dawson, Virginia P. *Engines and Innovation: Lewis Laboratory and American Propulsion Technology.* Document SP-4306. Washington, D.C.: National Aeronautics and Space Administration, 1991.

Divine, Robert A. *The Sputnik Challenge: Eisenhower's Response to the Soviet Satellite.* New York: Oxford University Press, 1993.

Dornberger, Walter R. "The German V–2." In *The History of Rocket Technology,* edited by Eugene M. Emme, 29–45. Detroit: Wayne State University Press, 1964.

Dukes, Paul. *The Last Great Game.* New York: St. Martin's Press, 1989.

Dupree, A. Hunter. *Science in the Federal Government: A History of Policies and Activities.* Cambridge, Harvard University Press, Belknap Press, 1957; Baltimore: Johns Hopkins University Press, 1986.

Dyer, Davis. *TRW: Pioneering Technology and Innovation Since 1900.* Boston: Harvard Business School Press, 1998.

Dyer, Lee, and Gary D. Paulson. *Project Management, An Annotated Bibliography.* Ithaca: New York State School of Industrial and Labor Relations, Cornell University, 1976.

Edge, D. O., and J. N. Wolfe, eds. *Meaning and Control: Essays in Social Aspects of Science and Technology.* London: Tavistock Publications, 1973.

Edwards, Paul N. *The Closed World: Computers and the Politics of Discourse in Cold War America.* Cambridge, Mass.: MIT Press, 1996.

Emme, Eugene M., ed. *The History of Rocket Technology.* Detroit: Wayne State University Press, 1964.

Evangelista, Matthew. *Innovation and the Arms Race.* Ithaca, N.Y.: Cornell University Press, 1988.

Fagan, M. D., ed. *A History of Engineering and Science in the Bell System: War and Peace in the National Service (1925–1975),* vol. 2. Murray Hill, N.J.: Bell Telephone Laboratories, Inc., 1978.

Fallows, James. *National Defense.* New York: Vintage Books, 1981.

Fisher, Gene E., and Warren E. Waler. "Operations Research and the RAND Corporation." In *Encyclopedia of Operations Research & Management Science,* edited by Saul I. Gass and Carl M. Harris. Boston: Kluwer Academic, 1996.

Flamm, Kenneth. *Creating the Computer: Government, Industry, and High Technology.* Washington, D.C.: Brookings Institution, 1988.

Freeman, Eva C., ed. *MIT Lincoln Laboratory, Technology in the National Interest.* Lexington, Mass.: Lincoln Laboratory, MIT, 1995.

Frisbee, John L., ed. *Makers of the United States Air Force.* Washington, D.C.: Office of Air Force History, 1987.

Futrell, Robert Frank. *Ideas, Concepts, Doctrine: Basic Thinking in the United States Air Force 1907–1960.* Maxwell AFB, Ala.: Air University Press, 1989.

Gass, Saul I., and Carl M. Harris, eds. *Encyclopedia of Operations Research & Management Science.* Boston: Kluwer Academic, 1996.

Geiger, Roger L. *Research and Relevant Knowledge: American Universities Since World War II.* New York: Oxford University Press, 1993.

George, Claude S. *The History of Management Thought.* Englewood Cliffs, N.J.: Prentice Hall, 1972.

Getting, Ivan. *All in a Lifetime: Science in the Defense of Democracy.* New York: Vantage Press, 1989.

Goldschmidt, Bertrand. *The Atomic Complex, A Worldwide Political History of Nuclear Energy.* Translated by Bruce M. Adkins from *Le Complexe Atomique.* La Grange Park, Ill.: American Nuclear Society, 1982.

Goldstine, Herman H. *The Computer from Pascal to von Neumann.* Princeton, N.J.: Princeton University Press, 1972.

Gorn, Michael H. *Harnessing the Genie: Science and Technology Forecasting for the Air Force 1944–1986.* Washington, D.C.: Office of Air Force History, 1988.

———. *Vulcan's Forge: The Making of an Air Force Command for Weapons Acquisition (1950–1985)* Washington, D.C.: Office of History, Air Force Systems Command, 1989.

———, ed. *Prophecy Fulfilled: "Toward New Horizons" and Its Legacy.* Washington, D.C.: Air Force History and Museums Program, 1994.

Green, Constance McLaughlin. *Vanguard—A History.* Document SP-4202. Washington, D.C.: National Aeronautics and Space Administration, 1970.

Hagen, John P. "The Viking and the Vanguard." In *The History of Rocket Technology,* edited by Eugene M. Emme, 122–41. Detroit: Wayne State University Press, 1964.

Hall, R. Cargill. "Early U.S. Satellite Proposals." In *The History of Rocket Technology,* edited by Eugene M. Emme, 67–106. Detroit: Wayne State University Press, 1964.

———. "The Eisenhower Administration and the Cold War: Framing American Astronautics to Serve National Security." In *Organizing for the Use of Space: Historical Perspectives on a Persistent Issue,* edited by Roger D. Launius, 49–61. American Astronautical Society History Series, vol. 18, ed. R. Cargill Hall. San Diego, Calif.: Univelt, Inc., 1995.

Hallion, Richard P. *On the Frontier: Flight Research at Dryden, 1946–1981.* Document SP-4303. Washington, D.C.: National Aeronautics and Space Administration, 1984.

Hansen, James R. *Engineer in Charge: A History of the Langley Aeronautical Laboratory, 1917–1958.* Document SP-4305. Washington, D.C.: National Aeronautics and Space Administration, 1987.

Hart, David M. *Forged Consensus: Science, Technology and Economic Policy in the United States, 1921–1953.* Princeton, N.J.: Princeton University Press, 1998.

Harwood, William B. *Raise Heaven and Earth: The Story of Martin Marietta People and Their Pioneering Accomplishments.* New York: Simon and Schuster, 1993.

Heims, Steve J. *John von Neumann and Norbert Wiener: From Mathematics to the Technologies of Life and Death.* Cambridge: MIT Press, 1980.

———. *The Cybernetics Group.* Cambridge: MIT Press, 1991.

Heppenheimer, T. A. *Countdown: A History of Space Flight.* New York: John Wiley and Sons, 1997.

Hewes, James E. *From Root to McNamara: Army Organization and Administration, 1900–1963.* Washington, D.C.: Center of Military History, United States Army, 1975.

Hoddeson, Lillian, Paul W. Henriksen, Roger A. Meade, and Catherine Westfall. *Critical Assembly: A Technical History of Los Alamos during the Oppenheimer Years, 1943–1945.* Cambridge: Cambridge University Press, 1993.

Holley, Irving Brinton, Jr. *Buying Aircraft: Matériel Procurement for the Army Air Forces.* Vol. 7 of *United States Army in World War II,* edited by Stetson Conn. Washington, D.C.: Office of the Chief of Military History, Department of the Army, 1964.

Hoos, Ida R. *Systems Analysis in Public Policy: A Critique.* Berkeley: University of California Press, 1972.

Hughes, Thomas P. "The Evolution of Large Technological Systems." In *The Social Construction of Technological Systems,* edited by Wiebe E. Bijker, Thomas P. Hughes, and Trevor Pinch, 51–82. Cambridge: MIT Press, 1987.

———. *American Genesis: A Century of Invention and Technological Enthusiasm.* New York: Viking Penguin, 1989.

———. *Rescuing Prometheus.* New York: Pantheon Books, 1998.

Hunley, J. D. "A Question of Antecedents: Peenemünde, JPL, and the Launching of U.S. Rocketry." In *Organizing for the Use of Space: Historical Perspectives on a Persistent Issue,* edited by Roger D. Launius, 1–31. American Astronautical Society History Series, vol. 18, ed. R. Cargill Hall. San Diego, Calif.: Univelt, Inc., 1995.

Jacobs, John F. *The SAGE Air Defense System: A Personal History.* Bedford, Mass.: MITRE Corporation, 1986.

Joerges, Bernward. "Large Technical Systems: Concepts and Issues." In *The Development of Large Technical Systems,* edited by Renate Mayntz and Thomas P. Hughes, 9–36. Boulder, Colo.: Westview Press, 1988.

Johnson, Stephen B. "From Concurrency to Phased Planning: An Episode in the History of Systems Management." In *Systems, Experts, and Computers,* edited by Agatha C. Hughes and Thomas P. Hughes. Chicago: University of Chicago Press, forthcoming.

Jones, Vincent C. *Manhattan: The Army and the Atomic Bomb.* Washington, D.C.: Center of Military History, United States Army, 1985.

Kaplan, Fred. *The Wizards of Armageddon.* Stanford, Calif.: Stanford University Press, 1983.

Kaplan, Norman, ed. *Science and Society.* Chicago: Rand McNally, 1965.

Kaufman, Allen. "In the Procurement Officer We Trust: Constitutional Norms, Air Force Procurement and Industrial Organization, 1938–1948." Working paper, MIT Defense and Arms Control Studies Program, Massachusetts Institute of Technology, Cambridge, 1996.

Kevles, Daniel J. *The Physicists: The History of a Scientific Community in Modern America.* Cambridge: Harvard University Press, 1971.

Kleinknecht, Kenneth S. "The Rocket Research Airplanes." In *The History of Rocket Technology,* edited by Eugene M. Emme, 189–211. Detroit: Wayne State University Press, 1964.

Kleinman, Daniel Lee. *Politics on the Endless Frontier: Postwar Research Policy in the United States.* Durham, N.C.: Duke University Press, 1995.

Komons, Nick A. *Science and the Air Force: A History of the Air Force Office of Scientific Research.* Arlington, Va.: Historical Division, Office of Information, Office of Aerospace Research, 1966.

Koppes, Clayton R. *JPL and the American Space Program: A History of the Jet Propulsion Laboratory.* New Haven, Conn.: Yale University Press, 1982.

Lafeber, Walter. *America, Russia, and the Cold War 1945–1992.* 7th ed. New York: McGraw-Hill, 1993.

Lamoreaux, Naomi R., and Daniel M. G. Raff, eds. *Coordination and Information: Historical Perspectives on the Organization of Enterprise.* Chicago: University of Chicago Press, 1995.

Launius, Roger D., ed. *Organizing for the Use of Space: Historical Perspectives on a Persistent Issue.* American Astronautical Society History Series, vol. 18, ed. R. Cargill Hall. San Diego, Calif.: Univelt, Inc., 1995.

Leslie, Stuart. *The Cold War and American Science.* New York: Columbia University Press, 1993.

Levine, Arnold S. *Managing NASA in the Apollo Era.* Document SP-4102. Washington, D.C.: National Aeronautics and Space Administration, 1982.

Lilienfeld, Robert. *The Rise of Systems Theory: An Ideological Analysis.* New York: John Wiley and Sons, 1978.

Lonnquest, John C., and David F. Winkler. *To Defend and Deter: The Legacy of the United States Cold War Missile Program.* Special report 97/01. Champaign, Ill.: United States Army Construction Engineering Research Laboratories, 1996.

Malina, Frank J. "Origins and First Decade of the Jet Propulsion Laboratory." In *The History of Rocket Technology,* edited by Eugene M. Emme, 46–66. Detroit: Wayne State University Press, 1964.

Mark, Hans, and Arnold Levine. *The Management of Research Institutions: A Look at Government Laboratories.* Document SP-481. Washington, D.C.: National Aeronautics and Space Administration, 1984.

Markusen, Ann, Scott Campbell, Peter Hall, and Sabina Deitrick. *The Rise of the Gunbelt: The Military Remapping of Industrial America.* Oxford: Oxford University Press, 1991.

Maurer, Maurer. *Aviation in the U.S. Army 1919–1939.* Washington, D.C.: Office of Air Force History, 1987.

Mayo-Wells, Wilfrid J. "The Origins of Space Telemetry." In *The History of Rocket Technology,* edited by Eugene M. Emme, 253–70. Detroit: Wayne State University Press, 1964.

McCurdy, Howard. *Inside NASA.* Baltimore: Johns Hopkins University Press, 1993.

McDougall, Walter A. *The Heavens and the Earth: A Political History of the Space Age.* New York: Basic Books, 1985.

McKenzie, Donald. *Inventing Accuracy: A Historical Sociology of Nuclear Missile Guidance.* Cambridge: MIT Press, 1990.

McNeill, William H. *The Pursuit of Power: Technology, Armed Force, and Society Since A.D. 1000.* Chicago: University of Chicago Press, 1982.

Miles, Wyndham D. "The Polaris." In *The History of Rocket Technology,* edited by Eugene M. Emme, 162–88. Detroit: Wayne State University Press, 1964.

MITRE Corporation. *MITRE: The First Twenty Years, A History of MITRE Corporation 1958–1978.* Bedford, Mass.: MITRE Corporation, 1979.

Mokyr, Joel. *The Lever of Riches: Technological Creativity and Economic Progress.* Oxford: Oxford University Press, 1990.

Moody, Walton S. *Building a Strategic Air Force.* Washington, D.C.: Air Force History and Museums Program, 1996.

Murphy, Howard R. *The Early History of the MITRE Corporation: Its Background, Inception, and First Five Years.* Document M72-110. Bedford, Mass.: MITRE Corporation, 1972.

Murray, Williamson, and Allan R. Millett, eds. *Military Innovation in the Interwar Period.* Cambridge: Cambridge University Press, 1996.

Neal, Roy. *Ace in the Hole: The Story of the Minuteman Missile.* Garden City, N.Y.: Doubleday and Company, 1962.

Nelson, Daniel. "Scientific Management in Retrospect." In *A Mental Revolution: Scientific Management Since Taylor,* edited by Daniel Nelson, 5–39. Columbus: Ohio State University Press, 1992.

———. "The Transformation of University Business Education." In *A Mental Revolution: Scientific Management Since Taylor,* edited by Daniel Nelson, 77–101. Columbus: Ohio State University Press, 1992.

————, ed. *A Mental Revolution: Scientific Management Since Taylor.* Columbus: Ohio State University Press, 1992.

Neufeld, Jacob. "Bernard A. Schriever: Challenging the Unknown." In *Makers of the United States Air Force,* edited by John L. Frisbee, 281–306. Washington, D.C.: Office of Air Force History, 1987.

————. *Ballistic Missiles in the United States Air Force 1945–1960.* Washington, D.C.: Office of Air Force History, 1990.

————, ed. *Reflections on Research and Development in the United States Air Force: An Interview with General Bernard A. Schriever and Generals Samuel C. Phillips, Robert T. Marsh, and James H. Doolittle, and Dr. Ivan A. Getting.* Interviews conducted by Richard Kohn. Washington, D.C.: Center for Air Force History, 1993.

Neufeld, Michael J. *The Rocket and the Reich: Peenemünde and the Coming of the Ballistic Missile Era.* New York: Free Press, 1995.

Newell, Homer E. *Beyond the Atmosphere, Early Years of Space Science.* Document SP-4211. Washington, D.C.: National Aeronautics and Space Administration, 1980.

Nieburg, H. L. *In the Name of Science.* Chicago: Quadrangle Books, 1966.

Noble, David F. *America by Design: Science, Technology, and the Rise of Corporate Capitalism.* New York: Alfred A. Knopf, 1977.

Norberg, Arthur L., and Judy E. O'Neill, with contributions by Kerry J. Freedman. *Transforming Computer Technology: Information Processing for the Pentagon, 1962–1986.* Baltimore: Johns Hopkins University Press, 1996.

Ordway, F. I. III, and M. Sharpe. *The Rocket Team.* New York: Thomas Y. Crowell, 1979.

Palmer, Gregory. *The McNamara Strategy and the Vietnam War: Program Budgeting in the Pentagon, 1960–1968.* Westport, Conn.: Greenwood Press, 1978.

Parker, R. H. *Managerial Accounting: An Historical Perspective.* London: Macmillan, 1969.

Penrose, Edith. *The Theory of the Growth of the Firm.* Oxford: Basil Blackwell, 1959.

Perrow, Charles. *Normal Accidents.* New York: Basic Books, 1984.

Perry, Robert L. "The Atlas, Thor, Titan, and Minuteman." In *The History of Rocket Technology,* edited by Eugene M. Emme, 142–61. Detroit: Wayne State University Press, 1964.

Petroski, Henry. *To Engineer Is Human: The Role of Failure in Successful Design.* New York: Vintage Books, 1982.

Porter, Michael. *Competitive Advantage of Nations.* New York: Free Press, 1990.

Porter, Theodore M. *Trust in Numbers: The Pursuit of Objectivity in Science and Public Life.* Princeton: Princeton University Press, 1995.

Pugh, Emerson W. *Memories That Shaped an Industry.* Cambridge: MIT Press, 1984.

————. *Building IBM: Shaping an Industry and Its Technology.* Cambridge: MIT Press, 1995.

Redmond, Kent C., and Thomas M. Smith. *Project Whirlwind Case History.* Manuscript ed. Bedford, Mass.: MITRE Corporation, 1975.

Reich, Leonard. *The Making of American Industrial Research: Science and Business at GE and Bell, 1876–1926.* Cambridge: Cambridge University Press, 1985.

Rhodes, Richard. *The Making of the Atomic Bomb*. New York: Simon and Schuster, Touchstone, 1986.

———. *Dark Sun: The Making of the Hydrogen Bomb*. New York: Simon and Schuster, Touchstone, 1995.

Rider, Robin E. "Operations Research and Game Theory: Early Connections." In *Toward a History of Game Theory*, edited by Roy E. Weintraub, 225–39. Annual supplement to *History of Political Economy*, vol. 24. Durham, N.C.: Duke University Press, 1992.

Roherty, James M. *Decisions of Robert S. McNamara*. Coral Gables, Fla.: University of Miami Press, 1970.

Roland, Alex. *Model Research, The National Advisory Committee for Aeronautics, 1915–1958*. Document SP-4103. Washington, D.C.: National Aeronautics and Space Administration, 1985.

Rosenberg, Nathan. *Inside the Black Box: Technology and Economics*. Cambridge: Cambridge University Press, 1982.

———. *Exploring the Black Box: Technology, Economics, and History*. Cambridge: Cambridge University Press, 1994.

Rycroft, Michael, ed. *The Cambridge Encyclopedia of Space*. Cambridge: Cambridge University Press, 1990.

Sapolsky, Harvey M. *The Polaris System Development, Bureaucratic and Programmatic Success in Government*. Cambridge: Harvard University Press, 1972.

———. "Myth and Reality in Project Planning and Control." In *Macro-Engineering and the Future, A Management Perspective*, edited by Frank P. Davidson and C. Lawrence Meador, 173–82. Boulder, Colo.: Westview Press, 1982.

———. *Science and the Navy: The History of the Office of Naval Research*. Princeton, N.J.: Princeton University Press, 1990.

Schaffel, Kenneth. *The Emerging Shield: The Air Force and the Evolution of Continental Air Defense, 1945–1960*. Washington, D.C.: Office of Air Force History, 1991.

Schwiebert, Ernest G. *A History of the U.S. Air Force Ballistic Missiles*. New York: Frederick A. Praeger, 1964.

Seidman, Harold, and Robert Gilmour. *Politics, Position, and Power: From the Positive to the Regulatory State*. 4th ed. New York: Oxford University Press, 1986.

Shapley, Deborah. *Promise and Power, The Life and Times of Robert McNamara*. Boston: Little, Brown and Company, 1993.

Sherry, Michael S. *The Rise of American Air Power: The Creation of Armageddon*. New Haven, Conn.: Yale University Press, 1987.

Shiman, Philip. *Forging the Sword: Defense Production During the Cold War*. Special report 97/77. Champaign, Ill.: United States Army Construction Engineering Research Laboratories, 1997.

Smith, Bruce L. R. *The RAND Corporation*. Cambridge: Harvard University Press, 1966.

———. *American Science Policy Since World War II*. Washington, D.C.: Brookings Institution, 1990.

Spanardi, Graham. *From Polaris to Trident: The Development of U.S. Fleet Ballistic Missile Technology*. New York: Cambridge University Press, 1994.

Spires, David. N. *Beyond Horizons: A Half Century of Air Force Space Leadership.* Washington, D.C.: U.S. Government Printing Office, 1997.

Stanley, Dennis J., and John J. Weaver. *An Air Force Command for R&D, 1949–1976: The History of ARDC/AFSC.* Washington, D.C.: Office of History, Headquarters, Air Force Systems Command, 1976.

Sturdevant, Rick W. "The United States Air Force Organizes for Space: The Operational Quest." In *Organizing for the Use of Space: Historical Perspectives on a Persistent Issue,* edited by Roger D. Launius, 155–86. American Astronautical Society History Series, vol. 18, ed. R. Cargill Hall. San Diego, Calif.: Univelt, Inc., 1995.

Sturm, Thomas A. *The USAF Scientific Advisory Board: Its First Twenty Years 1944–1964.* Washington, D.C.: USAF Historical Division Liaison Office, 1967.

Thomson, Ross, ed. *Learning and Technological Change.* New York: St. Martin's Press, 1993.

Trist, Eric. "A Socio-Technical Critique of Scientific Management." In *Meaning and Control: Essays in Social Aspects of Science and Technology,* edited by D. O. Edge and J. N. Wolfe, 95–116. London: Tavistock Publications, 1973.

Van Creveld, Martin. *Technology and War: From 2000 B.C. to the Present.* New York: Free Press, 1989.

Van Nostrand, A. D. *Fundable Knowledge: The Marketing of Defense Technology.* Mahwah, N.J.: Lawrence Erlbaum Associates, 1997.

Vander Muelen, Jacob. *Building the B–29.* Washington, D.C.: Smithsonian Institution Press, 1995.

Vincenti, Walter G. *What Engineers Know and How They Know It: Analytical Studies from Aerospace History.* Baltimore: Johns Hopkins University Press, 1990.

Von Braun, Wernher, and Frederick Ordway III. *History of Rocketry and Space Travel.* New York: Thomas Y. Crowell, 1966.

———. *The Rocket's Red Glare.* Garden City, N.Y.: Anchor Press, 1976.

Waring, Stephen P. *Taylorism Transformed: Scientific Management Theory Since 1945.* Chapel Hill: University of North Carolina Press, 1991.

———. "Peter Drucker, MBO, and the Corporatist Critique of Scientific Management." In *A Mental Revolution: Scientific Management Since Taylor,* edited by Daniel Nelson, 205–36. Columbus: Ohio State University Press, 1992.

Weintraub, Roy E., ed. *Toward a History of Game Theory.* Annual supplement to vol. 24 of *History of Political Economy.* Durham, N.C.: Duke University Press, 1992.

Wildes, Karl L., and Nilo A. Lindgren. *A Century of Electrical Engineering and Computer Science at MIT, 1882–1982.* Cambridge: MIT Press, 1985.

Winter, Frank H. *Rockets into Space.* Cambridge: Harvard University Press, 1990.

Wise, George. *Willis R. Whitney, General Electric, and the Origins of U.S. Industrial Research.* New York: Columbia University Press, 1985.

Witze, Claude. "The USAF Missile Program: A Management Milestone." In *A History of the U.S. Air Force Ballistic Missiles,* edited by Ernest G. Schwiebert, 167–83. New York: Frederick A. Praeger, 1964.

Wolk, Herman S. *Planning and Organizing the Postwar Air Force, 1943–1947.* Washington, D.C.: Office of Air Force History, 1984.

Wren, Daniel A. *The Evolution of Management Thought.* 2d ed. New York: John Wiley and Sons, 1979.

Yates, JoAnne. *Control Through Communication: The Rise of System in American Management.* Baltimore: Johns Hopkins University Press, 1989.

Articles and Dissertations

Akera, Atsushi. "Calculating a Natural World: Scientists, Engineers, and Computers in the United States, 1937–1968." Ph.D. diss., University of Pennsylvania, 1998.

Astrahan, Morton M., and John F. Jacobs. "History of the Design of the SAGE Computer—The AN/FSQ-7." *Annals of the History of Computing* 5, no. 4 (October 1983): 340–49.

Baldwin, Carliss Y., and Kim B. Clark. "Capital-Budgeting Systems and Capabilities Investments in U.S. Companies after the Second World War." *Business History Review* 68 (spring 1994): 73–109.

Baucom, Donald Ralph. "Air Force Images of Research and Development and Their Reflections in Organizational Structure and Management Policies," Ph.D. diss., University of Oklahoma, 1976.

Benington, Herbert D. "Production of Large Computer Programs." *Annals of the History of Computing* 5, no. 4 (1983): 350–61.

Bugos, Glenn E. "Manufacturing Certainty: Testing and Program Management for the F–4 Phantom II." *Social Studies of Science* 23 (1993): 265–300.

Cohen, M. D., R. Burkhart, G. Dosi, M. Egidi, L. Marengo, M. Warglien, and S. Winter. "Routines and Other Recurring Action Patterns of Organizations: Contemporary Research Issues." *Industrial and Corporate Change* 5, no. 3 (1996): 653–98.

Cowan, R., and D. Foray. "The Economics of Codification and the Diffusion of Knowledge." *Industrial and Corporate Change* 6, no. 3 (1997): 595–622.

Dennis, Michael Aaron. "'Our First Line of Defense': Two University Laboratories in the Postwar American State." *Isis* 85, no. 3 (1994): 427–55.

Everett, Robert R., Charles A. Zraket, and Herbert D. Benington. "SAGE—A Data-Processing System for Air Defense." *Annals of the History of Computing* 5, no. 4 (1983): 330–39.

Forrester, J. W. "Reliability of Components." *Annals of the History of Computing* 5, no. 4 (1983): 399–400.

Fortun, Michael, and Sylvan S. Schweber. "Scientists and the Legacy of World War II: The Case of Operations Research (OR)." *Social Studies of Science* 23 (1993): 595–642.

Galison, Peter. "The Ontology of the Enemy: Norbert Wiener and the Cybernetic Vision." *Critical Inquiry* 21 (fall 1994): 228–66.

Haig, Lt. Col. Thomas O. "Systems Engineering and Technical Direction as a Management Concept." Student research report no. 67. Washington, D.C.: Industrial College of the Armed Forces, 1966.

Harrington, John V. "Radar Data Transmission." *Annals of the History of Computing* 5, no. 4 (1983): 370–74.

Hounshell, David A. "Hughesian History of Technology and Chandlerian Business History: Parallels, Departures, and Critics." *History and Technology* 12 (1995): 205–24.

———. "The Medium Is the Message, or How Context Matters: The RAND Corporation Builds an Economics of Innovation, 1946–1962." Paper presented at the

Symposium on the Spread of the Systems Approach, Dibner Institute, Cambridge, Mass., May 5, 1996.

———. "The Cold War, RAND, and the Generation of Knowledge, 1946–1962." *Historical Studies in the Physical and Biological Sciences* 27, part 2 (1997): 237–68.

Howell, Charles. "Toward a History of Management Thought." *Business and Economic History* 24, no. 1 (1995): 41–58.

Jacobs, John F. "SAGE Overview." *Annals of the History of Computing* 5, no. 4 (1983): 323–29.

Jardini, David R. "Out of the Blue Yonder: The RAND Corporation's Diversification into Social Welfare Research, 1946–1968." Ph.D. diss., Carnegie Mellon University, 1996.

Joerges, Bernward. "Große technische Systeme: Zum Problem technischer Größenordnung und Maßstäblichkeit. *Technik und Gesellschaft* 6 (1992): 41–72.

Johns, Claude J., Jr. "The United States Air Force Intercontinental Ballistic Missile Program, 1954–1959: Technological Change and Organizational Innovation." Ph.D. diss., University of North Carolina, Chapel Hill, 1964.

Johnson, Stephen B. "Insuring the Future: The Development and Diffusion of Systems Management in the American and European Space Programs." Ph.D. diss., University of Minnesota, 1997.

———. "Three Approaches to Big Technology: Operations Research, Systems Engineering, and Project Management." *Technology and Culture* 38, no. 4 (October 1997): 891–919.

Layton, Edwin T. "Through the Looking Glass; or News from Lake Mirror Image." *Technology and Culture* 29 (1987): 594–607.

Lonnquest, John. "The Face of Atlas: General Bernard Schriever and the Development of the Atlas Intercontinental Ballistic Missile 1953–1960." Ph.D. diss., Duke University, 1996.

———. "The Evolution of Systems Management and Concurrency in the Air Force ICBM Program, 1947–1963." Paper presented at the Society for Military History conference, Montgomery, Ala., April 11, 1997.

McKenna, Christopher D. "The Origins of Management Consulting." *Business and Economic History* 24, no. 1 (1995): 51–58.

Mindell, David A. "Automation's Finest Hour: Radar and System Integration in World War II." Paper presented at the Symposium on the Spread of the Systems Approach, Dibner Institute, Cambridge, Mass., May 5, 1996.

———. "Datum for Its Own Annihilation: Feedback, Control, and Computing, 1916–1945." Ph.D. diss., Massachusetts Institute of Technology, 1996.

Nelson, Richard R., and Gavin Wright. "The Rise and Fall of American Technological Leadership: The Postwar Era in Perspective." *Journal of Economic Literature* 30 (December 1992): 1931–64.

———. "Technical Change as Cultural Evolution." In *Learning and Technological Change,* edited by Ross Thomson, 9–23. New York: St. Martin's Press, 1993.

Norberg, Arthur L. "High-Technology Calculation in the Early 20th Century: Punched Card Machinery in Business and Government." *Technology and Culture* 31, no. 4 (1990): 753–79.

Penrose, Edith. "The Growth of the Firm—A Case Study: The Hercules Powder Company." *Business History Review* 34 (spring 1960): 1–34.

Rau, Erik P. "New Times, New Uses: Philip Morse, the Cold War, and the Proliferation of Operations Research." Paper presented at the Symposium on the Spread of the Systems Approach, Dibner Institute, Cambridge, Mass., May 5, 1996.

Redmond, Kent C., and Thomas M. Smith. "From Whirlwind to MITRE: The R&D Story of the SAGE Air Defense Computer." MITRE Archives, 1997.

Rees, Mina. "The Computing Program of the Office of Naval Research, 1946–1953." *Communications of the ACM* 30, no. 10 (October 1987): 831–47.

Rosenberg, Nathan. "Technological Interdependence in the American Economy." *Technology and Culture* 19 (1979): 25–50.

Sapolsky, Harvey M., Eugene Gholz, and Allen Kaufman. "Security Lessons from the Cold War." *Foreign Affairs* 78, no. 4 (1999): 77–89.

Shepard, Herbert A. "Nine Dilemmas in Industrial Research." *Administrative Science Quarterly* 1, no. 2 (1956): 295–309.

Sigethy, Robert. "The Air Force Organization for Basic Research 1945–1970: A Study in Change." Ph.D. diss., American University, 1980.

Smith, Thomas M. "Project Whirlwind: An Unorthodox Development Project." *Technology and Culture* 17, no. 3 (1976): 447–64.

Sobsczac, Thomas V. "Network Planning—A Bibliography." *Journal of Industrial Engineering* 13, no. 6 (November–December 1962).

Thomson, Ross. "The Firm and Technological Change: From Managerial Capitalism to Nineteenth-Century Innovation and Back Again." *Business and Economic History* 22, no. 2 (winter 1993): 99–134.

Tropp, Henry S. "A Perspective on SAGE: Discussion." *Annals of the History of Computing* 5, no. 4 (1983): 375–98.

———. "SAGE at North Bay." *Annals of the History of Computing* 5, no. 4 (1983): 401–2.

Valley, George E. "How the SAGE Development Began." *Annals of the History of Computing* 7, no. 3 (1985): 196–226.

Waring, Stephen P. "Cold Calculus: The Cold War and Operations Research." *Radical History Review* 63 (fall 1995): 28–51.

Wieser, C. Robert. "The Cape Cod System." *Annals of the History of Computing* 5, no. 4 (1983): 362–69.

Index